Positive Duties to Wild Animals

This book further develops the interventionist literature on wild animal suffering using different theoretical frameworks, including some that have never previously been used to ground our positive duties to wild animals.

Though we've always known that the wild is a nasty place where predators lethally attack prey, only recently have most animal ethicists come to realize that most wild animals fail to flourish. In fact, what we know about wild animal reproduction suggests that the majority of sentient beings born into the world may not even live lives worth living. It's not unreasonable for one to initially respond to the above with a sense of depressed resignation, but a growing number of ethicists believe that we both can and should intervene. The purpose of this book is to further develop the interventionist literature by bringing together philosophers who agree that we have significant duties to help wild animals, but who use different theoretical frameworks, or who disagree about the details, e.g., about the reasons that ground our obligations to help wild animals, about how those obligations should be classified, about the content of our obligations, about the means we should use to fulfill our obligations, etc.

This book will be an invaluable resource for scholars, researchers and students of animal ethics, animal welfare, environmental ethics, philosophy, and sustainability. It was originally published as a special issue of the journal *Ethics, Policy & Environment*.

Kyle Johannsen is Sessional Faculty Member in the Department of Philosophy at Trent University, and an Animals in Philosophy, Politics, Law, and Ethics (APPLE) fellow at Queen's University. He's also a podcast host on the New Books Network's Animal Studies channel.

Routledge Environmental Ethics

Series Editor: Benjamin Hale, University of Colorado, Boulder

The Routledge Environmental Ethics series aims to gather novel work on questions that fall at the intersection of the normative and the practical, with an eye toward conceptual issues that bear on environmental policy and environmental science. Recognizing the growing need for input from academic philosophers and political theorists in the broader environmental discourse, but also acknowledging that moral responsibilities for environmental alteration cannot be understood without rooting themselves in the practical and descriptive details, this series aims to unify contributions from within the environmental literature.

Books in this series can cover topics in a range of environmental contexts, including individual responsibility for climate change, conceptual matters affecting climate policy, the moral underpinnings of endangered species protection, complications facing wildlife management, the nature of extinction, the ethics of reintroduction and assisted migration, reparative responsibilities to restore, among many others.

Climate Justice Beyond the State
Lachlan Umbers and Jeremy Moss

The Concept of Milieu in Environmental Ethics
Individual Responsibility within an Interconnected World
Laÿna Droz

Critical Realism and the Objective Value of Sustainability
Philosophical and Ethical Approaches
Gabriela-Lucia Sabau

Ecological Ethics and the Philosophy of Simone Weil
Decreation for the Anthropocene
Kathryn Lawson

Virtue Ethics and the Environment
Dominika Dzwonkowska

Positive Duties to Wild Animals
Kyle Johannsen

For more information on the series, please visit
https://www.routledge.com/Routledge-Environmental-Ethics/book-series/ENVE

Positive Duties to Wild Animals

Edited by
Kyle Johannsen

Routledge
Taylor & Francis Group
LONDON AND NEW YORK

First published 2025
by Routledge
4 Park Square, Milton Park, Abingdon, Oxon OX14 4RN

and by Routledge
605 Third Avenue, New York, NY 10158

Routledge is an imprint of the Taylor & Francis Group, an informa business

British Library Cataloguing in Publication Data
A catalogue record for this book is available from the British Library

ISBN13: 978-1-032-89818-6 (hbk)
ISBN13: 978-1-032-89817-9 (pbk)
ISBN13: 978-1-003-54473-9 (ebk)

DOI: 10.4324/9781003544739

Typeset in Myriad Pro
by Newgen Publishing UK

Publisher's Note
The publisher accepts responsibility for any inconsistencies that may have arisen during the conversion of this book from journal articles to book chapters, namely the inclusion of journal terminology.

Contents

Citation Information

The chapters in this book were originally published in *Ethics, Policy & Environment*, volume 26 issue 2 (2023). When citing this material, please use the original page numbering for each article, as follows:

Chapter 1
Positive Duties to Wild Animals: Introduction
Kyle Johannsen
Ethics, Policy & Environment, volume 26 issue 2 (2023), pp. 153–158

Chapter 2
Wild Animal Ethics: A Freedom-Based Approach
Eze Paez
Ethics, Policy & Environment, volume 26 issue 2 (2023), pp. 159–178

Chapter 3
Vulnerability and the Ethics of Environmental Enhancement
Catia Faria
Ethics, Policy & Environment, volume 26 issue 2 (2023), pp. 179–197

Chapter 4
Solidarity with Wild Animals
Mara-Daria Cojocaru and Alasdair Cochrane
Ethics, Policy & Environment, volume 26 issue 2 (2023), pp. 198–216

Chapter 5
Reducing Wild Animal Suffering Effectively: Why Impracticability and Normative Objections Fail Against the Most Promising Ways of Helping Wild Animals
Oscar Horta and Dayron Teran
Ethics, Policy & Environment, volume 26 issue 2 (2023), pp. 217–230

Chapter 6
Welcoming, Wild Animals, and Obligations to Assist
Josh Milburn
Ethics, Policy & Environment, volume 26 issue 2 (2023), pp. 231–248

Chapter 7

The Rebugnant Conclusion: Utilitarianism, Insects, Microbes, and AI Systems
Jeff Sebo
Ethics, Policy & Environment, volume 26 issue 2 (2023), pp. 249–264

Notes on Contributors

Alasdair Cochrane, Department of Politics and International Relations, University of Sheffield, Sheffield, UK.

Mara-Daria Cojocaru, Munich School of Philosophy, Munich, Germany.

Catia Faria, Department of Philosophy and Society, Complutense University of Madrid, Spain.

Oscar Horta, Department of Philosophy and Anthropology, University of Santiago de Compostela, Santiago de Compostela, Spain.

Kyle Johannsen, Department of Philosophy, Trent University, Peterborough, ON, Canada.

Josh Milburn, International Relations, Politics and History, Loughborough University, Loughborough, UK.

Eze Paez, Law and Philosophy Group, Pompeu Fabra University, Barcelona, Spain.

Jeff Sebo, Department of Environmental Studies, New York University, NY, USA.

Dayron Teran, Department of Philosophy and Anthropology, University of Santiago de Compostela, Santiago de Compostela, Spain.

Positive Duties to Wild Animals: Introduction

Kyle Johannsen

ABSTRACT
Though we've always known that the wild is a nasty place where predators lethally attack prey, only recently have most animal ethicists come to realize that most wild animals fail to flourish. In fact, what we know about wild animal reproduction suggests that the majority of sentient beings born into the world may not even live lives worth living. It's not unreasonable for one to initially respond to the above with a sense of depressed resignation, but a growing number ethicists believe that we both can and should intervene. The purpose of this special issue is to further develop the interventionist literature by bringing togeher authors who agree that we owe significant postive duties to wild animals, but who use different theoretical frameworks, or who disagree about the details, e.g., about the reasons that ground our positive obligations to wild animals, about how those positive obligations should be classififed, about the content of our positive obligations, about the means we should use to fulfill our positive obligations, etc.

Though we've always known that the wild is a nasty place where predators lethally attack prey, only recently have most animal ethicists come to realize that most wild animals fail to flourish. In fact, what we know about wild animal reproduction suggests that the majority of sentient beings born into the world may not even live lives worth living. After all, only some wild animals (K-Strategists) protect their genes by devoting considerable energy to each of their offspring. Many animals protect their genes by instead producing large numbers of offspring. This evolutionary reproductive strategy, normally referred to as the 'r-Strategy' (MacArthur & Wilson, 1967; Pianka, 1970), is used by many lizards, amphibians, fish and small mammals. Since r-Strategists' offspring individually receive little energy and are too numerous for ecosystems to support, the majority of those offspring end up dying from disease, starvation, injury, exposure or predation, shortly after birth.

The r-Strategy is not the only source of suffering in nature. This seems obvious enough when we note that other causes, such as predation or food scarcity, must be present in order for r-Strategists' young to die prematurely, but it's also true that r-Strategist infants aren't the only wild animals who experience a low level of welfare. Most (sentient) K-Strategist animals and r-Strategist adults endure a considerable amount of suffering from a variety of sources. Consider the effects of predation. In addition to the painful deaths caused by predation, the threat of predation causes hunger by preventing prey animals from foraging in risky areas (Anholt & Werner, 1995; McNamara & Houston, 1987), it's a source of psychological stress for prey animals (Creel et al., 2007; Dwyer, 2004), and predation is a source of non-fatal injury for those who manage to escape predator attacks.

It's not unreasonable for one to initially respond to the above with a sense of depressed resignation, but a growing number of ethicists believe that we both can and should intervene. Some interventions, specifically large-scale ones, will require research before they can be conducted safely and effectively, but the sheer scale of wild animal suffering means we have strong moral reasons to fund that research. According to a recent estimate, the world's population of wild, terrestrial vertebrates is about one trillion, give or take a 0, and the number of wild marine vertebrates is even larger (Tomasik, 2019). As large as the above estimate is, though, we should keep in mind that it specifically represents the total number of (wild, terrestrial) vertebrates alive at any given moment. As a result, it doesn't include the number of vertebrates who, over a period of time, e.g. 10 years, were born and then died shortly afterward. Were we to include the many r-Strategist vertebrates who die a painful death only shortly after being born, our population number would be far higher than one trillion.

Though it seems 'natural' to think of intervention in the wild as a matter of humanitarian assistance (Duclos, 2022; Faria, 2023; Horta, 2017; Johannsen, 2020, 2021; McMahan, 2010, 2015; Paez, 2015), a number of different understandings have emerged in recent years. For example, a number of writers argue that cosmopolitan distributive justice militates in favor of intervention (Cochrane, 2018; Faria, 2014; Horta, 2016). After all, the core intuition driving cosmopolitanism is that the circumstances of one's birth, including one's nationality, are morally arbitrary. Accordingly, cosmopolitans maintain that it's unfair when one's nationality determines one's life prospects. But if we judge it unfair for the circumstances of birth to determine one's life prospects, doesn't that judgment include species membership, too? The species one happens to be born into is just as circumstantial as the nation one's born in. And if inequalities traceable to species membership are unfair, then wild animals are in a tremendously unfair situation. On average, wild animals are far worse off than human beings. In light of this inequality, do we not have reasons of distributive justice to intervene?

In addition, a number of authors have argued that we owe rectification to wild animals for anthropogenic harms (Palmer, 2021; Pepper, 2019; Sebo, 2021, 2022). Clare Palmer, for example, highlights various anthropogenic harms that she thinks ground duties of rectification, but among the most significant harms are those associated with anthropogenic climate change. In her words, 'Changing precipitation patterns and intensity, rising temperatures, warming of the upper oceans, Arctic ice melt, sea level rise, heat waves, the shifting of habitat types, will all impact wild animal populations. While some populations will be able to take advantage of the changing conditions to expand and grow, others will be severely affected, including many already threatened and endangered species (Palmer, 2021, pp. 179–180)'. Considering how pervasive the effects of climate change both are and will be, it seems that not only distributive, but also rectificatory justice, militates in favor of a considerable level of intervention in the wild. Forms of rectificatory intervention she considers are rescue and rehabilitation, habitat restoration, and assisted migration (Palmer, 2021, pp. 185–192).[1]

The purpose of this special issue is to further develop the interventionist literature by bringing together authors who agree that we owe significant positive duties to wild animals, but who use different theoretical frameworks, or who disagree about the details, e.g. about the reasons that ground our positive obligations to wild animals, about how those positive obligations should be classified, about the content of our positive obligations, about the means we should use to fulfill our positive obligations, etc. Indeed, some of the papers in this collection use frameworks that have never previously been used to ground our positive duties to wild animals.

In 'Wild Animal Ethics: A Freedom-based Approach', Eze Paez claims that, contrary to what some have argued (Donaldson & Kymlicka, 2011), respect for wild animals' freedom is not in tension with intervening to help wild animals. Quite the opposite: respect for wild animals' freedom actually entails that we have an enforceable, political duty to provide them with assistance. According to Paez, there are two main dimensions, or components, of freedom. The first – security - is a matter of being secure against control-undermining interference. The second – richness - is a matter of how rich one's choice situations are, i.e. the number and quality of options one has to choose from. At present, most wild animals' options are quite limited: they can choose to fight or flee; to look for food and risk predation, or to hide and risk starvation; etc. In light of the limitations on wild animals' options, Paez argues that respect for their freedom requires that we intervene to enrich their choice situations.

In 'Vulnerability and the Ethics of Environmental Enhancement', Catia Faria analyzes intervention via the framework provided by Mackenzie et al. (2014). Though some authors have rightly suggested that wild animals are vulnerable to harms caused by human beings, such as the harms associated with anthropogenic climate change (Palmer, 2021), Faria argues that wild animals are vulnerable to much more than just us. She notes that most wild animals die prematurely from natural causes, so even animals who live within well-functioning ecosystems are in a highly vulnerable situation. In other words, many wild animals within even well-functioning ecosystems are currently vulnerable – they're presently unable to meet their basic needs. What's more, the fact that, for the most part, we aren't currently bothering to assist wild animals, means that they're dispositionally vulnerable, too, i.e. that even those wild animals who are able to satisfy their needs are at high risk of being unable to do so in the future. Faria concludes that since we have a duty to assist vulnerable populations, we also have a duty to assist wild animals. Of particular interest to Faria are interventions that assist sentient wild animals by modifying their environment, rather than by modifying sentient wild animals themselves.

In 'Solidarity with Wild Animals', Mara-Daria Cojocaru and Alasdair Cochrane approach wild animal ethics through the lens of Sally Scholz's work on solidarity (Scholz, 2008). Though it seems clear enough (upon reflection) that solidarity obtains between us and domesticated animals, the possibility that it obtains between us and wild animals has thus far been largely unexplored. According to Cojocaru and Cochrane, there are both descriptive and normative senses of solidarity that obtain between us and wild animals. Descriptively speaking, it's true that, as a matter of fact, we have social relationships with many wild animals, and that we feel a sense of community with those animals. It's also true that wild animals are oppressed in various ways, whether it be in circuses and zoos, or via the harmful effects of anthropogenic climate change. Those who seek to liberate wild animals from circuses and zoos, or to rehabilitate animals harmed by the effects of climate change, are responding to a perfectly appropriate, normative conviction. Finally, it's true that the sort of institutions we associate with the welfare state extend in certain (very limited) ways to wild animals, e.g. via public health initiatives like the One Health initiative, or via publicly funded conservation initiatives. In light of the fact that social relationships already exist between human beings and wild animals, and that some of the vulnerabilities wild animals possess are caused by us, Cojocaru and Cochrane argue that we have good reason to further develop and extend the institutional support we provide them (to further extend civic solidarity).

The collection's remaining papers concern practical and theoretical questions that emerge when we begin to think seriously about our positive duties to wild animals. Perhaps the most immediate practical question is: Which interventions will effectively improve wild animals' welfare? In 'Reducing Wild Animal Suffering Effectively: Why Impracticability and Normative Objections Fail Against the Most Promising Ways of Helping Wild Animals', Oscar Horta and Dayron Teran discuss a series of intervention strategies that they think are worthy of further development. Of particular interest are three areas of research that scientists consider to be especially promising (Animal Ethics, 2020): wild animal vaccination programs, interventions that aim to assist wild animals affected by extreme weather events, and interventions that aim to assist wild animals who live in urban and suburban environments. In addition, the authors claim that contraception programs, and incorporating wild animal welfare into conservation-related policies and initiatives, are promising as well. Though there are various objections to both the feasibility and desirability of assisting wild animals, Horta and Terán argue that these objections are ineffective against the above strategies for improving wild animal welfare.

In 'Welcoming, Wild Animals, and Obligations to Assist', Josh Milburn investigates the issue of what we owe to wild animals who we've welcomed into our spaces, such as pigeons who are fed in urban spaces, or farmland animals who have become the focus of conservation efforts. Though Milburn is skeptical that we have a general duty to assist wild animals, he does think that we sometimes owe special duties of assistance to wild animals in cases where we've formed morally significant relationships with them. One such relationship, he notes, is the relationship we form with animals who we welcome into our spaces, and who in turn become dependent upon the resources we've made available. Milburn argues that an upshot of this relationship is that we have at least a *prima facie* reason to oppose the rewilding of some human spaces. For example, it may be the case that increased rates of veganism, and the development of vertical agriculture, will make some existing farmland unnecessary to us. If this happens, we should keep in mind that the land may still be needed by the wild animals who live on it and who, if they were deliberately encouraged, we have a morally significant relationship with. Perhaps at least some human spaces should be 'conserved' for the sake of the wild animals we've invited into them, since those spaces have effectively become their habitat.

In 'The Rebugnant Conclusion: Insects, Microbes, Aggregation, and Fanaticism', Jeff Sebo applies two (arguably problematic) implications of utilitarianism to the context of wild animal population ethics. One is the well-known 'Repugnant Conclusion', i.e. the conclusion that we ought to increase the size of human populations, even when doing so decreases quality of life, because increasing the size of populations increases aggregate welfare (Parfit, 1984, pp. 381–390). Sebo notes that a similar 'Re*bug*nant Conclusion' applies to wild animals: given the choice between a small number of large animals who each possess a high amount of welfare, and a much larger numbers of small animals who each possess a low (but still net positive) amount of welfare, it seems that we should prefer the latter, since such populations have higher aggregate welfare. The second sort of implication Sebo discusses is associated with a thought experiment called 'Pascal's Mugging' (Bostrom, 2009). The issue here is that maximizing expected value seems to require taking courses of action that, if successful, would yield astronomically large gains, even when the chances of success are astronomically low. Sebo notes that a similar 'Pascal's *Bug*ging' applies to wild animals: maximizing expected value seems to require prioritizing the interests of astronomically large populations (such as populations of microbes), even if those populations' members have an incredibly low chance of being sentient, and even if each individual would possess an incredibly low capacity for welfare if they were sentient. Though Sebo argues that the Re*bug*nant Conclusion and Pascal's *Bug*ging should be accepted in theory, he also argues that these conclusions may be avoidable in practice.

Note

1. Much of the content in the above paragraphs was taken from Johannsen (2021).

Disclosure statement

No potential conflict of interest was reported by the author(s).

ORCID

Kyle Johannsen http://orcid.org/0000-0002-8169-9719

References

Anholt, B. R., & Werner, E. E. (1995). Interaction between food availability and predation mortality mediated by adaptive behavior. *Ecology*, *76*(7), 2230–2234. https://doi.org/10.2307/1941696

Animal Ethics. (2020). Surveying attitudes towards helping wild animals among scientists and students. *Animal Ethics*. Retrieved September 19, 2022, from https://www.animal-ethics.org/survey-helping-wild-animals-scientists-students/

Bostrom, N. (2009). Pascal's wager. *Analysis*, *69*(3), 443–445. https://doi.org/10.1093/analys/anp062

Cochrane, A. (2018). *Sentientist politics: A theory of global inter-species justice*. Oxford University Press.

Creel, S., Liley, S., Christianson, D., & Winnie, J. A., Jr. (2007). Predation risk affects reproductive physiology and demography of elk. *Science*, *315*(5814), 960. https://doi.org/10.1126/science.1135918

Donaldson, S., & Kymlicka, W. (2011). *Zoopolis: A political theory of animal rights*. Oxford University Press.

Duclos, J. (2022). *Wilderness, morality, and value*. Lexington Books.

Dwyer, C. M. (2004). How has the risk of predation shaped the behavioural responses of sheep to fear and distress? *Animal Welfare*, *13*(3), 269–281. https://doi.org/10.1017/S0962728600028384

Faria, C. (2014). Equality, priority and non-human animals. *Dilemata*, *14*, 225–236.

Faria, C. (2023). *Animal ethics in the wild: Wild animal suffering and intervention in nature*. Cambridge University Press.

Horta, O. (2016). Egalitarianism and animals. *Between the Species, 19*(1), 108–144.
Horta, O. (2017). Animal suffering in nature: The case for intervention. *Environmental Ethics, 39*(3), 261–279. https://doi.org/10.5840/enviroethics201739320
Johannsen, K. (2020). To assist or not to assist? Assessing the potential moral costs of humanitarian intervention in nature. *Environmental Values, 29*(1), 29–45. https://doi.org/10.3197/096327119X15579936382644
Johannsen, K. (2021). *Wild animal ethics: The moral and political problem of wild animal suffering*. Routledge.
MacArthur, R. H., & Wilson, E. O. (1967). *The theory of island biogeography*. Princeton University Press.
Mackenzie, C., Rogers, W., & Dodds, S. (Eds.). (2014). *Vulnerability: New essays in ethics and feminist philosophy*. Oxford University Press.
McMahan, J. (2010). The meat eaters. *The New York Times*.
McMahan, J. (2015). The moral problem of predation. In A. Chignell, T. Cuneo, & M. Halteman (Eds.), *Philosophy comes to dinner: Arguments about the ethics of eating* (pp. 268–294). Routledge.
McNamara, J. M., & Houston, A. I. (1987). Starvation and predation as factors limiting population size. *Ecology, 68*(5), 1515–1519. https://doi.org/10.2307/1939235
Paez, E. (2015). Refusing help and inflicting harm: A critique of the environmentalist view. *Relations, 3*, 165–178. https://doi.org/10.7358/rela-2015-002-paez
Palmer, C. (2021). Assisting wild animals vulnerable to climate change: Why ethical strategies diverge. *Journal of Applied Philosophy, 38*(2), 179–195. https://doi.org/10.1111/japp.12358
Parfit, D. (1984). *Reasons and persons*. Oxford University Press.
Pepper, A. (2019). Adapting to climate change: What we owe to other animals. *Journal of Applied Philosophy, 36*(4), 592–607. https://doi.org/10.1111/japp.12337
Pianka, E. R. (1970). On r- and K-Selection. *The American Naturalist, 104*(940), 592–597. https://doi.org/10.1086/282697
Scholz, S. (2008). *Political solidarity*. Pennsylvania State University Press.
Sebo, J. (2021). Animals and climate change. In M. Budolfson, T. McPherson, & D. Plunkett (Eds.), *Philosophy and climate change* (pp. 42–66). Oxford University Press.
Sebo, J. (2022). *Saving animals, saving ourselves: Why animals matter for pandemics, climate change, and other catastrophes*. Oxford University Press.
Tomasik, B. (2019). How many wild animals are there? *Essays on Reducing Suffering*. Retrieved March 29, 2022, from https://reducing-suffering.org/how-many-wild-animals-are-there/

Wild Animals Ethics: A Freedom-Based Approach

Eze Paez

ABSTRACT
On expectation, most wild animals have lives of net suffering due to naturogenic causes. Some have claimed that concern for their well-being gives us reasons to intervene in nature on their behalf. Against this, it has been said that many interventions to assist wild animals would be wrong, even if successful, because they would violate their freedom. According to the Freedom-based Approach I defend in this paper, this view is misguided. Concern for wild animal freedom does indeed gives us reasons to secure these animals against control-undermining interferences, but also to intervene in nature in order to enrich their choices.

On some estimates, around one quintillion nonhuman animals live in the wild (Tomasik, 2009). They routinely endure serious harms inflicted not by human action, but by the ordinary workings of nature. Starvation, thirst, disease, parasites, extreme weather conditions, aggressions by other animals, and the reproductive strategy followed by a majority of wild animal species are examples of those harmful processes and events.[1] There are, I think, sufficient reasons to believe that, on expectation, wild animals have lives of net suffering due to naturogenic causes, and I will assume as much throughout this paper.[2] I will also assume that sentient animals merit a robust moral and political standing. Thus, this is a problem to be taken very seriously.

In this article I will defend what may be called a Freedom-based Approach to wild animal ethics, according to which:

(a) Wild animal freedom of choice should be adopted as a political ideal, that is, a guide for the design of our political institutions and a source of enforceable obligations;
(b) These obligations are mainly of two sorts: securing wild animals against control-undermining interferences and intervening in nature in order to enrich, to an extent, their fundamental choice-types.

My reasons for advancing a Freedom-based Approach are twofold. First, the dominant approach in the wild animal ethics literature is welfarist. According to this view, our reasons to intervene in nature in order to assist wild animals are derived from the value of their well-being. This welfarist strategy has been employed to defend an array of different positions – from the mere permissibility of intervening in nature (Palmer, 2010), to the existence of enforceable duties of assistance (Cochrane, 2018).[3]

Second, nonhuman animal freedom has usually been invoked as a source of reasons *against* intervening in nature on behalf of animals. Most notably, Sue Donaldson and Will Kymlicka have argued that human interventions directed at eliminating the risks posed by nature involve a 'terrible curtailment' of wild animals' individual and collective freedom (Donaldson & Kymlicka, 2011, p. 166). Contrariwise, I think that if we are to take the political ideal of wild animal freedom seriously, protecting animals against human interferences that prevent them from living their lives on their own terms is necessary, but not sufficient. We also have a duty to create 'an adequate range of options' for our fellow agents to choose from (Raz, 1986, p. 408). Thus, to the extent that wild animal freedom is reason-giving, it gives reasons for, rather than against, intervening in nature to assist wild animals.

Unfortunately, I do not have the space to provide a fully developed treatment of the Freedom-based Approach. I have the more modest aim of defending the claim that the ideal of freedom is applicable to wild animals and that we can derive from this ideal an option-enrichment obligation on their behalf. If my argument is sound it would follow, then, that freedom-based and welfarist approaches to wild animal ethics are not mutually exclusive. It would be possible to see both well-being and freedom of choice as equally valid, though separate, grounds for an obligation to intervene in nature.

I will start by discussing how the value of an agent's freedom of choice provides us with reasons both to secure them against control-undermining interferences and to enrich their choice-types (Section 1). I will subsequently provide an outline of our freedom-based obligations toward wild animals (Section 2). Then I will deploy two different strategies in support of the claim that wild animal freedom is a source of enforceable obligations. On the one hand, I will argue, against Alasdair Cochrane and Robert Garner, that if the freedom of typical human adults is intrinsically valuable, and a source of such obligations, then so is that of other agents (Section 3). On the other, I will explain why views that deny that freedom matters at all in any fundamental way – views such as welfare-consequentialism – should nevertheless accept freedom as a political ideal (Section 4). Afterward, I will address two major objections to my view. One claims that free human-nonhuman animal relations are impossible, since animals cannot deliberate with us about the principles that should govern our interactions (Section 5). Another, derived from Donaldson's and Kymlicka's views, claims that we are generally required not to intervene in nature, since option-enriching interferences undermine wild animal autonomy (Section 6). I will end with some concluding remarks (Section 7).

One final clarification is in order, however. My views are meant to apply solely to those nonhuman animals who qualify as intentional agents. As will become apparent over the course of my argument, it seems to be true that all sentient animals (animals with a capacity for affective mental states, such as pleasure or pain) possess intentional agency (Wilcox, 2019). The scientific consensus seems to be that all vertebrate animals are sentient (Low, 2012), there being also substantial evidence of sentience in cephalopods and decapods (Birch et al., 2021).

1. Freedom of Choice

For the purposes of my argument, I will understand freedom as a property of choices. There are several ideals we can refer to, however, when we predicate freedom of an agent's choice. Here I am concerned with freedom as an ideal of social relations.[4] A complete account of that ideal ought to specify the conditions that a relationship between agents must possess if one of them is to preserve control over their choices even when the other has the capacity to interfere with them. I will be assuming throughout my discussion that freedom in our social relations should be regarded as a political value, that is, as I said, as a regulative ideal for institutional design and as the source of some of our enforceable obligations toward others. This is what I will call social, or political, freedom.[5]

Like all agents, wild animals face choice-situations, that is, scenarios in which they must decide among two or more options. Thus, they must consider the different alternatives available to them regarding, for instance, where to build a nest or a den, where to forage or hunt, where to drink or pasture, or how to distribute scarce resources among offspring. For wild animals, as for any other agent, not all kinds of choice-situations – what hereafter I shall call choice-types – are equally important. There is a subset of those choice-types over which an agent must exercise freedom of choice if their life is to be valuable for them, by their own lights.[6] Our main concern ought to be directed at guaranteeing agents a certain degree of freedom regarding such fundamental choice-types.

The likely prevalence of naturogenic harms in wild animals' lives implies that, in these and other fundamental choice-types, most animals must opt between a limited number of very poor alternatives. Consider, for example, the choice between, on the one hand, increasing the risk of death by dehydration or starvation and, on the other, increasing the risk of being devoured when trying to alleviate thirst in a predator-infested waterhole (Faria, 2016, p. 68). Because of the ordinary workings of nature, wild animals' existence qua agents consists in a string of dramatic choices: fleeing from danger or confronting it (including fighting their attackers), a more or a less probable death, or experiencing greater or lesser suffering.

Let us put all of this in a more formal way. Consider a wild animal, *W*, and a particular choice-situation, *C*, defined by a set of mutually exclusive and jointly exhaustive options, *x* and *y*. Now we will say that

W's choice is free with respect to *C (x, y)* to the extent that:

(i) *W* has the capacity to choose among any of the options, *x* and *y*, that define *C*, and
(ii) No other agent interferes with that choice in ways that undermine *W*'s control over it

W's freedom with regards to *C* may be increased in two different ways. One consists in making *W*'s freedom more *secure* against interferences that undermine their control over their choices. The other consists in *enriching* their freedom by improving the options contained in the choice situations they face (cf. Raz, 1986, pp. 407–408).

Consider, first, that dimension of freedom we may call its *security*. *W*'s freedom of choice with respect to *C* is compromised if other agents interfere with them in ways which undermine, totally or in part, *W*'s control over their choice. Therefore, freedom in a certain choice-situation requires not only the capacity to select the options that define it, but also, in the last instance, that it is the agent's will that is effective in the exercise of choice, as opposed to the alien will of another agent. The degree to which *W*'s freedom with respect to *C* is secure will depend on the extent to which she is immune from being interfered with by other agents in ways that can be understood to undermine their control over *C*.

How are we to identify such control-undermining interferences? Different theories of social freedom will advance more or less expansive conceptions of these. For instance, they may disagree about whether interference always requires changing the defining options of a choice-situation (like removing one of them or replacing it with a worse alternative) or whether it can also consist in manipulating an agent into falsely believing that any of these structural elements have changed. They can also disagree about the modal status interferences must have. Some may claim that, for freedom in a choice-situation to count as secure, absence of interference with *W*'s actual choice is enough (Hobbes, 1651/2008). Others may additionally require absence of interference in any of the possible choices available to *W*, both actual and counterfactual (Berlin, 1969/1990; Carter, 1999; Kramer, 2003). According to this position, *W* is secure in their freedom of choice to the extent that they are not interfered with, no matter what they happen to choose. Some may even endorse the more demanding view that security in our freedom of choice further requires absence of interference even when the potentially interfering party finds *W*'s choice uncongenial. Proponents of this view would say that we are secure in our freedom of choice only if there are institutional guardrails that deprive other agents of their *power* of interference with us, even if they do not intend to make use of it, or that guarantee that such interference can only occur with our permission (Pettit, 1997). Nevertheless, whatever our preferred account of social freedom, keeping all else constant, *W* enjoys greater freedom with respect to *C* the more secure they are against the relevant kinds of interference.[7]

Now, just as an agent's freedom in a choice-situation may be more or less secure, it can be more or less *rich*. This is that aspect of freedom which is variously called in the literature the 'adequacy of choice' (Raz, 1986, p. 373) or 'opportunity-freedom' (Taylor, 1979). A choice-situation, as we understand it here, is defined by the specific options which appear *ex ante* to the agent as genuinely open for choice, as opposed to unavailable for them, prior to any considerations about the possible interference of others. What determines which specific options are *ex ante* open in this way are the resources the agent possesses. These include their physical and cognitive capabilities – including the ability to identify whether a certain option is genuinely open for choice – the conditions of their social and natural environment, the material resources required to access the option or (at least in the case of most humans) the risk of incurring some social penalty by doing so. If an agent lacks the individual, natural and social resources necessary for accessing an option, then that option is ineligible for them. It is not part of the actual choice-situation they are facing.

The richer the range of alternatives available for an agent in a given choice-situation, all else being the same, the greater their freedom of choice is. Suppose *W*'s security in their control over *C* is kept constant. For any area of choice (say, nesting, foraging, hunting, or playing), depending on the resources at their disposal, an agent can be presented with any number of eligible options. It is possible to enrich *W*'s freedom of choice with respect to *C* by augmenting the quantity of alternatives contained in the choice-set, thereby adding more options for *W* to choose from. This would entail changing *C* (*x*, *y*) into *C* (*x*, *y*, . . . , *n*). However, if such a change in quantity is to count as an enrichment, not any addition will do. First, a choice-set is not enriched by the introduction of trivial alternatives or alternatives with potentially horrendous consequences (Raz, 1986, p. 374). The newly added options must be significant from the agent's perspective, allowing them to achieve a more efficient or granular satisfaction of their aims than was the case with the previous array of alternatives at their disposal.[8] Second, the choice-set should not be dominated by the need to protect the agent's life, where a choice is dominated in this way 'if all options except one will make the continuation of the life [the agent] has rather unlikely' (Raz, 1986, p. 376). Suppose, for example, that you were stranded on a small island with a predator who is continually hunting for you. In this scenario, your willpower and physical resources would be completely spent in your struggle for survival, so that you would be left with no opportunity 'to do or even to think of anything other than how to escape from the beast' (Raz, 1986, p. 374). At least in these cases, *removing* an option may result in the improvement of the choice-set, because it is thereby rendered undominated by the agent's need to protect their life.

I have argued so far that the value of an agent's freedom provides us with two kinds of reasons. The first are reasons to secure them against control-undermining interferences with their fundamental choice-types. The second are reasons to enrich those choice-types. In the following section I will discuss the option-enrichment obligation in greater detail.

2. The Freedom-Based Obligation to Enrich Wild Animal Choices

The claim that the ideal of freedom grounds an enforceable obligation to enrich, to some extent, the fundamental choice-types of agents, is not strange in the literature – when applied to human beings. It is a common trait of broadly-speaking egalitarian theories of justice – such as left-libertarianism (Otsuka, 2003), liberal egalitarianism (R. Dworkin, 2000; Rawls, 1971), or republicanism (Pettit, 1997). On the assumption that this ideal of freedom is applicable to nonhuman animals, it would follow that we also have a similar obligation toward them, including those animals that live in the wild. In this section I will start by discussing the general features of that obligation (Section 2.1). Then I will focus on specific kinds of enrichment interventions (Section 2.2).

2.1. General Features of the Option-Enrichment Obligation

The content of a freedom-based obligation to intervene in nature is different from that derived from welfarist principles. A welfarist ideal requires us to guarantee some particular *outcome* for wild animals with respect to their well-being. For instance, ensuring that these animals have maximally good lives (Pearce, 1995/2015). Contrariwise, the ideal of freedom is not outcome-oriented in this way. It requires us, first, to ensure that wild animals will be able to live their lives on their own terms; and, second, to add non-trivial options to their fundamental choice-types, as well as to remove life-threatening or otherwise potentially horrendous alternatives.[9] None of this implies, however, that our duty to enrich wild animals' choices is narrow in scope. It is not restricted to merely 'policing nature', that is, preventing predation (Cowen, 2003). It extends, rather, to all naturogenic harms which impoverish wild animals' fundamental choice-types.

Certainly, different versions of the freedom-based approach may specify this duty in different ways. For instance, we could claim that we have enforceable obligations to wild animals as fellow citizens (Cochrane, 2018, pp. 79–99). Alternatively, we could accept, in part, Donaldson & Kymlicka's view about the sovereignty of wild animal communities, but insist that justice in international relations requires us to perform option-enriching interventions (Donaldson & Kymlicka, 2011, pp. 156–186; see Section 6).

Moreover, different versions of the approach can assign us more or less demanding aims. Some accounts may tell us that we ought to provide wild animals with the *best* possible options in their basic choice-types. Some others could tell us, alternatively, that we must merely provide them with *reasonably good* alternatives. Note, however, that given the probable predominance of naturogenic harms in wild animals' lives, even this relatively modest aim would give us reasons to intervene in nature in quite expansive ways.

Finally, we need to distinguish between those cases in which a proposed enrichment intervention is Pareto optimal and those in which it is not. In the first kind of case, there will be a net improvement to the options in some wild animals' choice-types, without any resulting net impoverishment to the choice-types of other agents. When an enrichment is optimal in this way, it is compatible with the pursuit of a good life, as they may see it, by all agents involved, either because it does not affect them or because it improves their sphere of control over that pursuit. Thus, our reasons to enrich wild animal freedom will not contradict our reasons to secure it.

In some other cases, however, there will be those who are harmed by an enrichment intervention. That is, wild animals whose freedom of choice experiences a net impoverishment. Given that such interventions would be incompatible with their terms of choice, security establishes some limits on the permissibility of option-enriching interventions. A fully developed freedom-based account should provide us with criteria to determine, in these cases, which enrichment policies are justified and which are not.

2.2. Particular Kinds of Enrichment Interventions

Let us consider, first, *environmental enhancement*, or in other words, interventions aimed at reshaping non-sentient natural entities and processes (Faria, 2016, p. 144). Many of the non-anthropogenic harms suffered by wild animals are caused by non-sentient nature. This is the case with starvation, thirst, extreme weather conditions, disease, and non-sentient parasites and parasitoids. We could greatly enrich wild animals' fundamental choice-types by providing them with food, water, shelter and medical care. One proposal along these lines consists in extending the institutions of the welfare state to wild animals (Pearce, 2015). More modestly, it has been suggested that we may preserve members of all animal species 'in large wildlife parks where the worst forms of suffering can be avoided' (Moen, 2016, p. 100).

Secondly, our interventions could target sentient animals themselves, rather than their environment. Wild animals routinely interfere in the choices of others, blocking and replacing their options with worse alternatives, sometimes in an extremely aggressive way. Thus, some of our interventions might be aimed at minimizing such aggressive behavior.[10] For instance, we may be able to use gene-drive technology to reprogram predators into herbivores (McMahan, 2010; Pearce, 2009).[11] The aim of other interventions might be to enhance animals' psychological and physical capacities in ways that make new non-trivial options open for them or remove horrendous ones. Many animals reproduce by having enormous quantities of offspring, most of whom die shortly after coming into existence – in part due to competition for scarce resources. We may genetically redesign them to reduce their reproductive rate (Johannsen, 2017), thus allowing for a greater share of food for their progeny. In addition, we could make animals more resistant to disease or immune to parasites, better able to withstand harsh climatic conditions, or improve their efficiency in processing nutrients and water.[12]

Finally, we have a secondary obligation to improve our means of safely intervening in nature (Moen, 2016, p. 103). Many imaginable option-enriching interventions may turn out to be futile or have perverse effects. The permissibility of intervening in nature should be understood as conditioned by a proviso against futility and perversity (Faria, 2016, pp. 89–110). We should not be disposed to invest resources in option-enriching interventions with no positive impact or, indeed, which may cause harms greater than those they seek to mitigate. Moreover, permissible interventions may not be feasible, given our scientific knowledge and technological progress. Taking seriously the value of wild animal freedom should compel us to acknowledge that we also have freedom-derived reasons to increase the scope of the justifiable interventions we are capable of performing.

3. The Intrinsic Value of Wild Animal Freedom

In the previous sections, I merely assumed that wild animal freedom of choice is a source of enforceable obligations. I will now deploy my first argument to that effect, which is meant to address the views of authors such as Alasdair Cochrane or Robert Garner. They claim that the most important interests of sentient individuals ground enforceable duties, that is, duties which must be protected through the establishment of political institutions (Cochrane, 2018, pp. 26–35; Garner, 2013, pp. 93–94). They deny, however, that animals have an intrinsic interest in freedom which we should protect in this way (Cochrane, 2009, 2012, pp. 11–12; Garner, 2013, pp. 99–132).

My aim in this section is to counter the claim that animals lack an intrinsic interest in freedom. First, I explain how nonhuman animals are intentional agents and, therefore, the sort of entity which faces choice-situations (Section 2.1). Second, I argue that if the freedom of typical human adults is intrinsically valuable for them, then so is the freedom of other agents (Section 2.2). It follows then that, even on Cochrane's or Garner's views, wild animals' interest in freedom grounds enforceable obligations.

3.1. Nonhuman Animals are Intentional Agents

There is a growing body of literature recognizing that sentient nonhuman animals are intentional agents (Korsgaard, 2018; Rowlands, 2012, 2019; Sebo, 2017; Varner, 2012; Wilcox, 2019).[13] Like all such agents, sentient nonhuman animals function through the operations of two modules. First, a representational module which registers features of inner and outer states, on an evidential basis (Pettit, 1993). It allows them to represent the environment as being *such-and-such*, including the representation of certain options as open for them and as of the kind that must be pursued or that must be avoided. Second, a governing module, which conducts the agent's interactions with their environment on the basis of certain regularities. This module governs the agent in a reliable way, that is, causes them to abide by said regularities in a suitable range of possible worlds. For instance, if a certain feature of the environment *p* is registered as of the kind to be avoided, the agent will be disposed to display the behavior of avoiding *p*.

Admittedly, the registrations of nonhuman animal intentional agents do not take a linguistic form, but consist rather in the perception that something is the case (Glock, 2009). This also applies to those representations which, according to the rules of practical rationality, should elicit in the agent the motivation to act in certain ways. The representation that something is desirable need not take the form of a judgment about the reasons to desire it and to act in the appropriate ways. It can take the form of an 'affordance' or normative perceptual experience: the perception that *p* is *to-be-pursued* or *to-be-avoided*, which reliably generates the motivation to pursue or avoid *p*.[14] That is, the desire that *p*.

Therefore, animals are subjects with a capacity to choose (a) grounded in the possession of reliable evidence-based perceptual representations and (b) reliably governed by regularities of minimal theoretical and practical rationality. Animals entertain representations of their environment and their inner physiological states, including the perception of objects as avoidable and pursuable: painful or pleasurable, dangerous or friendly, disgusting or delicious, and so forth. Indeed, given the evidence available, the most plausible explanation of animal behavior is by appeal to the content of their intentional states (Dretske, 2006).[15]

3.2. *Animal Freedom as Intrinsically Valuable*

On a widespread view, an agent has the capacity for choice that matters for social freedom, if, and only if, she has certain robust metacognitive capacities.[16] These are the capacities that allow us to say that an agent is autonomous or possesses a free will. Such capacities include: the ability to entertain higher-order intentional states (Bratman, 1997; G. Dworkin, 1970; Frankfurt, 1971; Watson, 1975), or the ability to formulate, endorse and revise evaluative and normative standards in a reason-responsive way (Fischer, 1994; Fischer & Ravizza, 1998; Wolf, 1990). It appears that, unlike typical humans, most animals do not possess these sophisticated cognitive skills. According to this view, then, freedom of choice cannot be intrinsically valuable for nonhuman animals. I shall argue, however, that if the freedom of typical human agents is intrinsically valuable, then so is the freedom of other sentient agents.[17]

In order to understand the view, I am arguing against, it will be helpful to distinguish between three kinds of control which agents can have over their choices.[18] As intentional agents, animals can exercise *rational control*. That is, their beliefs and desires are updated in the face of new evidence and their actions are rationally responsive to those beliefs and desires. Maybe some (like great apes, elephants or cetaceans) have, in addition, *volitional control*. That is, they can form second-order desires which govern their actions. Still, all of them seem to lack the kind of control over their choices afforded by the possession of metacognitive capacities. We can call this kind of control *discursive control*. On the view

I am arguing against, the intrinsic value of freedom is grounded on the possession of this special kind of control – which, to the best of our knowledge, is only found among some human beings. Therefore, it could be said, the freedom of choice of other human and nonhuman agents cannot be intrinsically valuable for them. It can be, at most, *instrumentally* good, whenever choosing on their own terms happens to be most conducive to their well-being.

As I mentioned, this is the position which, in the animal ethics literature, has been adopted by Cochrane (2009, 2012) and Garner (2013). Several authors have argued, however, that it is a misguided view (Giroux, 2016; Paez, 2022; Wilcox, 2020). Suppose, as Wilcox claims, that the explanation for why most consider freedom of choice to be intrinsically valuable is that it consists in possessing 'the opportunity to determine the course of one's own life' (Wilcox, 2020, pp. 9, 11). In believing freedom is valuable in this way, we accept that there are *pro tanto* reasons to respect an agent's 'authority over their own choices and the direction of their lives' (Wilcox, 2020, p. 10).

As I explained above, intentional agents are constituted to form their beliefs and desires in accordance with certain standards of theoretical and practical rationality. These standards include the axiological and normative principles that determine, for each choice-situation, which option the agent should consider preferable and, therefore, set as the target for their effective disposition for action. These principles are, so to speak, the agent's own *terms of choice*. For an intentional agent to exercise control over a choice is for them to decide on such terms. That is, to decide according to the grounds which determine the choice-worthiness of the options they face in a choice-situation. Conversely, to choose on another agent's terms is to have their control over their choices undermined. In having their grounds of choice replaced by those of another agent, that aspect of their life becomes determined by another individual, rather than by herself. Her authority over their own choices is therefore usurped.

The above description of how an intentional agent's terms of choice determine the choice-worthiness of options is no less accurate when applied to agents without discursive control. Consider an agent that has rational control over their choices in the way we have described. Absent independent obstruction or perturbation, their beliefs and desires respond to the evidence she registers and their actions are a result of said beliefs and desires, all conforming to what we can identify as standards of theoretical and practical rationality. Moreover, this agent does not operate in this way because of sheer luck, but because of the representational and governing modules that make them the kind of reason-responsive agent she is. Certain events are perceived as inherently good or bad. These beliefs inform their desires. Together with other beliefs about the world, a strategy is chosen to satisfy those desires.

Thus, non-discursive intentional agents are also the kind of entities with a capacity to determine the course of their lives. If that is the ability that grounds the intrinsic value of freedom of choice in the case of typical humans, then freedom of choice must be similarly valuable for the other entities that possess it, including nonhuman animals (Wilcox, 2020, p. 16). Conversely, an agent's authority over a choice-situation is no less undermined by the imposition of another agent's terms simply because the former lacks the kind of control afforded by the possession of sophisticated cognitive capacities. Another individual's grounds of choice are substituted for hers. Their life becomes, at least in part, determined by an alien will.

If nonhuman animals have an intrinsic interest in freedom of choice, then, even on the view that an intrinsic interest is necessary for enforceability, animals' freedom can ground enforceable duties. Still, two clarifications are in order. First, reliably rational agents are not perfect. Their actual choices may not coincide with those that should follow from the standards implicit in their operations. This may give us reasons to assist them in ensuring the primacy of *their own terms*, thereby reinstalling their control over a choice-situation. The importance of restoring the primacy of one's own terms is grounds for justified paternalism toward children, some neuroatypical humans and nonhuman animals (Wilcox, 2020, p. 14). If, generally speaking, nonhuman animals are less reliable agents than typical humans, then the scope of justified paternalism will be greater for the former than for the latter.

Second, some may want to maintain the view that freedom of choice is more intrinsically valuable for discursive agents than for other kinds of agents. It could be said that discursive agents are like 'architects', with the ability to devise an overall shape for their lives, whereas non-discursive agents are like 'builders', who construct their lives piecemeal without following an overarching plan (Wilcox, 2020, pp. 15–16). We would therefore have, *ceteris paribus*, stronger reasons to secure and enrich the choice-types of typical human adults than those of non-discursive agents. Thankfully, the issue of whether this position is true does not affect the cogency of my overall argument, so I'm able to remain agnostic.

4. Freedom for Welfare-Consequentialists

What if, however, one denies that freedom of choice has any intrinsic value *at all*? This is indeed true of welfare-consequentialist views, which figure prominently in the animal ethics literature (Faria, 2014; Holtug, 2007; Horta, 2016; Singer, 1979/2011; Varner, 2012). In this section, I will argue that there are compelling reasons to believe, on welfare-consequentialist grounds, that freedom of choice (as expounded in Section 1) is one of the ideals to which our political institutions should conform. Thus, institutions should protect individuals' sphere of choice from interference, as well as enrich the options they have in certain choice-types.

It is common in the consequentialist literature to distinguish between our *criteria of rightness* and our *decision procedures*. The first identify the morally right and wrong actions in a certain choice. For instance, in the case of utilitarianism, the right course of action is whichever one maximizes well-being. Decision procedures, by contrast, are meant to help agents identify which of the actions available for them they should perform (Hare, 1981; Hooker, 2000, pp. 67–68).

This distinction is important for consequentialist theories, since there are a series of epistemic and motivational limitations that suggest that it may be self-defeating, or highly counterproductive, if agents tried to comply directly with consequentialist criteria of rightness. For instance, human agents operate with limited information, are not well-equipped to follow complex rules, suffer from biases and tend to be partial toward themselves and those they see as members of their group. Therefore, even if only well-being (and perhaps how it is distributed) matters in a fundamental way, our personal and institutional decision procedures should not simply tell us to choose whatever option or policy seems, on reflection, optimific.

Regarding our attitude toward agents' choices, welfare-consequentialists can admit that, overall, individuals are the best judge of their own interests (Mill, 1859/2001). According to a plausible interpretation of this principle, we should distinguish between the sphere of choice regarding which human beings are reliable rational agents and certain well-defined cases in which they are not (Goodin, 1995, p. 127). In general, typical human adults are likely to judge their interests 'more nearly correctly than anyone else' (Goodin, 1995, p. 127). This is compatible with the claim that, in some situations, individuals are poor judges of their interests. For instance, those cases involving adaptive preferences, probabilities, or the further future (Goodin, 1995, p. 128). Beyond these cases, however, human individuals can be considered reliable rational agents. Together with the epistemic and motivational limitations mentioned before, this suggests that in our relationships with others, we should adopt a general rule of respecting that sphere of choice over which individuals are competent, so that they may pursue their ends. The mere fact that we believe that it is optimific to interfere with other agents' choices would be insufficient reason to do so.

These considerations help explain why the ideal of freedom of choice might be the best workable approximation, in practice, to welfare-consequentialist ends. It tells us that, as a general rule, we ought to respect agents' choices in those domains in which they are competent. In addition, we ought to enrich the choice-situations in which they are to choose on their own terms, so as to allow them a more granular satisfaction of their aims.

From an institutional perspective, it follows that we ought to enforce a general duty to secure and enrich agents' freedom, rather than directly advance some welfarist end. Thus, 'policies and institutions should allow people to frame and pursue their own interests as they will' (Goodin, 1995, p. 131) in order to ensure in the long run that individuals have the best possible lives. Through these institutional mechanisms, such as rights, we not only allow agents to pursue their own goals, but also assign individuals a publicly recognized decision-making authority over their own life which others should respect (Woodard, 2019, pp. 129–130).

The institutional implications of welfare-consequentialism are similar for nonhuman animals. The epistemic and motivational limitations of human agents give us enough grounds for accepting that, on the ideal decision-making procedure, our relationships with nonhuman animals would also be subject to constraints against optimization (John & Sebo, 2020; Paez, 2019). In our dealings with them, we are also vulnerable to partiality and prone to making mistakes. Certainly, as mentioned above, the scope for justified paternalism might be greater in the case of animals. Still, regarding that sphere of choice in which animals are reliable rational agents, an equivalent of the best judge principle is applicable. Our institutions should be designed with the aim of securing that sphere of choice against unwarranted human interference, as well as with the aim of enriching the options from which animals can choose.[19]

5. The Alleged Impossibility of Free Human-Nonhuman Animal Relations

I have argued that our relations with animals, including wild animals, should be structured by the ideal of freedom of choice. I will now address the first of two objections to a Freedom-based Approach. Consider someone who endorses the view that nonhuman animal freedom is intrinsically valuable. They could object, in the following way, that free human-nonhuman animal relations are impossible:

> A relation between agents is free if, and only if, it is governed by principles which all parties in the relation would deem acceptable in ideal deliberative conditions. Nonhuman animals, however, lack the capacity to understand, formulate, and endorse normative principles. We cannot, therefore, reach a deliberative agreement with them about such principles. Thus, our relations with the other animals can only be governed by the terms *humans* impose and are, therefore, unfree for animals.

If this objection succeeded, it would follow that the ideal of freedom is not a useful guide for our relations with animals. No matter what we did, our interactions with them (including option-enriching interventions) would always undermine their control over their choices.

I think this is a misguided construal of what free relations with non-discursive intentional agents require. By extending Scanlon's (1998) trustee model to nonhuman animals, it is possible to explain how our relations with them can be compatible with their freedom. The correct governing principles for our relations with animals are simply those which cognitively sophisticated agents, acting as animals' trustees, identify as acceptable under ideal deliberative conditions, given nonhuman animals' terms of choice.

The objection assumes that a relation between agents is free only if it is ruled by commonly accepted principles. It must be conceded that, whenever common acceptance obtains, interactions between agents that comply with those principles cannot be said to be control-undermining for any of them. This is because each party can regard the terms on which the relation operates as *their own*. The objection, as formulated, is wrong to assume, however, that the existence of commonly agreed-upon principles is necessary for a relation between agents to be free. It is, rather, only sufficient.

Suppose that each party in a relationship desires to retain control over their choices despite the fact that the others can interfere with them. I advance that what is necessary in order to prevent their relation from being control-undermining is for them construct a *shared control* over their relationship. That shared control must be of the same kind each one possesses over their own choices. Otherwise, one or more of them suffers a usurpation of their control and, therefore, suffers unfreedom. Imagine, for instance, that the parties in question are typical human beings. Because the kind of control they possess is discursive (see Section 3.2), constructing a shared control of the same kind does indeed require the parties involved to adopt some principles in common. When their relationship is ruled by such commonly accepted principles each party retains their freedom of choice.

Yet nonhuman animals lack discursive control over their choices, possessing only rational or volitional control (see, again, Section 3.2). When considering what possessing a shared control over their relation would entail for nonhuman animals, there are no compelling reasons to require that it be based on the adoption, by them, of certain normative standards. That is the hallmark of discursive control. It is enough if the relationship is governed by terms compatible with both the rational or volitional control possessed by nonhuman animals and with the discursive control possessed by cognitively sophisticated agents. That is, terms which all parties could regard as their own in spite of being different kinds of intentional agents.

How, then, can we construct a shared human-nonhuman animal control over a relationship that satisfies these conditions? As I said above, here we can benefit from Scanlon's suggestion that we should act as the 'trustees' of cognitively unsophisticated humans in our deliberations, so that they can indirectly participate in it (Scanlon, 1998, p. 83). It has been argued that we should adopt this *trustee model* as well for our relations with animals (Burgis, 2018; Elmore, 2021; Talbert, 2006). The trustee model was introduced by Scanlon in order to accommodate the existence of obligations toward human individuals for whom things can go better or worse, but who cannot articulate reasoned objections regarding how metacognitively sophisticated agents treat them. Nonhuman sentient agents find themselves in a similar predicament. They are also individuals whose lives can go better or worse, and they cannot assess reasons or express objections. On the assumption that species-membership is irrelevant, there seems to be no compelling reasons why the trustee model should be applicable to non-discursive humans, but not to non-discursive animals.

It behooves discursive agents, as trustees of animals, to perform three main tasks. First, trustees should take nonhuman animals' perceptions of what is inherently avoidable and pursuable for them as, so to speak, raw data. Then trustees ought to deliberate, on the basis of this data, about which conceptions of nonhuman animals' good furnish a satisfactory articulation of those perceptions.[20] Finally, given those conceptions of the good, trustees should decide which principles are acceptable for relating with animals.

According to this trustee model, the shared governing principles that allow for free relations with nonhuman animals are simply those which discursive agents (acting as animals' trustees) identify as acceptable under ideal deliberative conditions, given nonhuman animals' terms of choice. Under such conditions, *ex hypothesi*, the conceptions of nonhuman animals' good which the trustees have articulated are those most compatible with the data. Moreover, under said conditions, the principles agreed upon to govern relations with animals would be premised on their full moral and political standing. Together, these two elements entail that the terms of the shared human-nonhuman animal relationship would be compatible with animals' rational and volitional control. In addition, under these ideal conditions, the terms of the relationship would be the result of a deliberative agreement among human discursive agents. Thus, they would be compatible with those agents' discursive control.

In this way, it is possible to construct a shared control over a human-nonhuman animal relationship in which the parties retain the same kind of control that each possesses as a separate intentional agent. Whenever shared control obtains, a relationship will be compatible with the terms of choice of both human discursive agents and nonhuman animals, enabling the latter to enjoy freedom of choice in spite of our capacity to interfere in their lives. In practice, our deliberation about the principles that ought to govern our relationships with animals will not take place in such optimal conditions. That is regrettable, so we must strive to achieve an ever closer approximation to the ideal of shared control.

6. Freedom-Derived Reasons Against Option-Enriching Interventions

Let us discuss one final objection to my Freedom-based Approach. Suppose, as I defend, that it is in principle possible to relate with animals in ways compatible with their freedom. Still, it could be said, what animal freedom requires of us is not to intervene in nature:

> We have freedom-derived reasons to enrich wild animals' choices, but our reasons not to undermine their freedom are generally stronger. Most option-enriching interventions are actually control-undermining. Therefore, we generally have decisive reasons against intervening in nature.

This is Sue Donaldson's and Will Kymlicka's view (Donaldson & Kymlicka, 2011, pp. 156–186). They assume that wild animals have an interest in their own individual and collective autonomy, since autonomy is taken to be a constitutive part of animals' flourishing (Donaldson & Kymlicka, 2011, p. 172). Donaldson and Kymlicka concede that there might be cases of permissible interventions which are not control-undermining. These are isolated small-scale interventions aimed at providing assistance to particular wild animals, or schemes to protect wild animals from events that pose an existential threat to their communities, the persistence of which is understood as a precondition for their autonomy (Donaldson & Kymlicka, 2011, pp. 179–180). Beyond these exceptional cases, however, our relations with wild animals and their communities should be governed by a general principle of nonintervention. I think, however, that we should reject this position.

Donaldson and Kymlicka provide three reasons in support of their view. The first is that many wild animals and their communities are already competent enough at self-regulating their affairs, despite the natural risks and harms amid which they live (Donaldson & Kymlicka, 2011, pp. 170–171). However, as some have already pointed out, that is not an accurate description of wild animals' situation. Wild animal communities are better understood as 'failed states', unable to guarantee the conditions that allow most of their members to survive and live free from deprivation (Cochrane, 2013; Horta, 2013).

Second, the authors admit that most wild animals may not be competent enough to navigate nature (Donaldson & Kymlicka, 2011, p. 177). Despite this admission, Donaldson and Kymlicka claim that we have reasons to endorse a general stance against option-enriching interventions. First, such interventions would allegedly undermine autonomy by causing wild animals to become dependent on permanent human assistance. Second, wild animal flourishing allegedly requires the exercise of those capacities which have been selected for by evolutionary pressures (Donaldson & Kymlicka, 2011, pp. 166, 182). Nevertheless, we should reject this argument. Animals are already dependent on their natural environment for the pursuit of their good, so making them dependent on human assistance would not create a new dependency relationship, but merely change one of the relata (Faria, 2016, pp. 143–144). Moreover, on this new form of dependency, unlike on the old one, animals would be provided with the resources needed to pursue a reasonably good life. Just as Donaldson and Kymlicka concede in the case of domesticated animals, if our relations of assistance are governed by terms compatible with animals' freedom, they would not be control-undermining (Donaldson & Kymlicka, 2011. pp. 84, 142–144). For instance, maybe we would have to ensure that option-enriching interventions do not depend on the contingent goodwill of human beings. Instead, those interventions ought to be part of the public policy of an interspecies democracy. For its part, 'flourishing' can refer to at least two different evaluative criteria – namely, whether an animal acts in accordance with their naturally evolved capacities or whether they successfully pursue their good (Faria, 2016, pp. 146–148). The second should be regarded as the relevant standard. Because of wild animals' impoverished conditions of choice, an appeal to their flourishing would give us reasons for, rather than against, enriching opportunities for the successful pursuit of their good.

Third, Donaldson and Kymlicka additionally suggest that wild animals have so intense an aversion to interacting with human beings that it generally trumps any alleged preference for less hostile natural entities and processes (Donaldson & Kymlicka, 2011, pp. 176–177).[21] Now, there might be some particular cases in which respecting an animal's terms requires us to abstain from interfering with either them or their environment.[22] However, what the authors advance is a conception of wild animals' good which posits a *generally* stronger aversion to benevolent human intervention than to natural harms. On the assumption that suffering likely prevails in nature, this seems implausible. Conditions in the wild are, on expectation, so poor that for most animals the quality of the options they face in their fundamental choice-types is very low – indeed, their options likely have negative prudential value. Enriching animals' choices, thereby providing them with more and better options in life, even at the cost of some unwanted interactions with humans, appears to be a generally acceptable trade-off. We ought to reject, as well, this argument against option-enriching interventions.

7. Conclusion

In this article, I have defended a novel foundation for the obligation to intervene in nature on behalf of wild animals. My main goal has been to show how the ideal of freedom of choice is applicable to wild animals, and how we can derive from it an enforceable obligation to enrich, to an extent, wild animals' basic choice-types.

Much else is needed in order to fully flesh out a Freedom-based Approach. Yet if I am right, my article's thesis is, I think, an important one. Some have worried that animal well-being and freedom are, so to speak, pitted against each other. No matter how benevolent or successful, our interventions to assist wild animals – or at least the most wide-ranging interventions – would infringe upon their freedom. If that were the case, we might justifiably regret the magnitude of the naturogenic harms wild animals suffer, but nevertheless believe that intervening in nature to assist them would be wrong. I have explained why I think this view is mistaken. Concern for wild animal freedom provides us with reasons for, rather than against, expansive intervention in nature.

Notes

1. Reproductive processes seem to be the most important source of naturogenic harms. The most widespread reproductive strategy, traditionally called *r*-selection, consists in having a high quantity of offspring in each reproductive cycle with very high juvenile mortality rates, such that, on average, only one hatchling per progenitor survives to maturity. To the extent

that we have reasons to believe that suffering predominates in nature, it is because of this strategy (Horta, 2010; Ng, 1995; Tomasik, 2015).

2. Let me briefly justify that assumption. Most wild animals are invertebrates. Evidence of their sentience is still scarce and the character of their experiences, such as they may possess, is not well known. Further research is needed in this respect (Groff & Ng, 2019). In the worst-case scenario, however, most invertebrate species would be sentient and capable of intense suffering. Since invertebrates are almost the totality of sentient beings, even if we assign low credence to this hypothesis, the amount of expected net suffering is still arguably very high.
3. See also: Nussbaum (2006); Pearce (1995/2015); Cowen (2003); McMahan (2015); Moen (2016); Faria (2016); Horta (2017); Paez (2020); and Johannsen (2020).
4. What are these other ideals about agents' choices? Freedom as a *psychological ideal* (the ideal of a free will) specifies the deliberative capacities which an agent must possess in order for them to have the kind of control over their choices which we consider necessary for moral accountability. For its part, freedom as an *ethical ideal* specifies the conditions for a reliable or excellent exercise of free will. For instance, not being a slave to one's passions. I shall not discuss this ideal further. I borrow these distinctions from Pettit (2015).
5. For the classic analysis of freedom of choice see MacCallum (1967).
6. I shall turn later, in Section 5, to the problems involved in interpreting nonhuman animals' good and identifying the principles that must govern our relations with them if our relations are to be compatible with their freedom.
7. For the purposes of my argument, however, we do not need to settle for any particular account of control-undermining interferences. All of them are compatible, I believe, with the view I defend here.
8. Note that adding further non-horrendous options can sometimes result in a worse choice-set. As one anonymous reviewer remarked, because agents are not perfectly rational, having more options may be anxiety-inducing for them. That can inhibit their ability to make a choice at all or, at least, cause them to feel overwhelmed, thus making a poorer choice than they would otherwise have.
9. Note that animals may enjoy sufficiently rich choice-situations and still possess a level of well-being below the threshold prescribed by some welfare-consequentialist ideal. This would indeed be regrettable from a welfare-consequentialist perspective. Nevertheless, if a freedom-based institutional design is the best working approximation of that ideal (Section 4), then welfare-consequentialists should endorse said design.
10. Two theoretical alternatives are open for us here. We can adopt the restrictive view that only intentional or negligent interferences by moral agents count as control-undermining. If so, interferences by other agents, including wild animals, can only impoverish freedom, but not compromise its security. Alternatively, we may claim that all agential interferences can be control-undermining, including those performed by nonhuman animals. If we believe that our reasons to secure an agents' freedom differ in strength from our reasons to enrich it, what view we adopt will make some difference.
11. Cowen suggests easing or removing the strictures against killing predators (Cowen, 2003, p. 182). Nevertheless, phasing out predators through genetic re-engineering appears to be more friendly to their terms of choice than killing them, even if the killing were done in a painless way. Genetic re-engineering would, therefore, be preferable. For the opposite view, however, see Bramble (2021).
12. Some worry that genetic re-engineering might not be identity-preserving. It would never result in a benefit for the re-designed individual, but rather in their replacement by a numerically different one. Thus, it would amount to killing some for the benefit of others, being, therefore, morally impermissible (Korsgaard, 2018, p. 187). I have argued elsewhere, however, that many important ways of re-designing animals would not face this problem (Paez, 2020).
13. See also Paez (2022) for an extended version of argument in this subsection.

14. The term 'affordances' was coined by Gibson (1979). For the role of normatively-loaded perceptions in animal intentionality, see Sebo (2017) and Korsgaard (2018).
15. Note, however, that none of this entails that animals are moral agents.
16. This is assumed in prominent accounts of political liberty. See, for instance, Rawls (1971, 1993); Nozick (1971); Pettit (1997); and Lovett (2010).
17. A less sophisticated version of this argument appeared in Paez (2022).
18. I borrow this classification of the different kinds of control from Pettit (2001, pp. 32–103).
19. More tentatively, there might be one further reason to adopt freedom as a regulative political ideal for animals, even if we are welfarists. It has been greatly advantageous to invoke that ideal in the efforts to emancipate those who had been denied equal political standing, including non-discursive humans. Therefore, at least in our present circumstances, it might be equally advantageous to employ freedom in the case of animals. Refusing to do so may reinforce the idea that nonhuman animals merit less than full political consideration.
20. For the need to act as interpreters of animals' good see Donaldson and Kymlicka (2011, pp. 108–112); Meijer (2019). Possibly, given any set of nonhuman aims there might be more than one conception of the good which is equally satisfactory at articulating them. If this is so, for any such animal, the set of plausible accounts of a life worth living for them remains underdetermined by their perceptions of what is inherently pursuable or avoidable.
21. I thank Will Kymlicka for stressing the need to address this point.
22. Note that, even in these cases, enriching wild animals' opportunities need not always involve them coming into contact with humans and that, if unavoidable, contact could be minimized. I thank an anonymous reviewer for suggesting this point.

Acknowledgments

I am especially thankful to Paula Casal, Alasdair Cochrane, Catia Faria, Valéry Giroux, Oscar Horta, François Jaquet, Kyle Johannsen, Will Kymlicka, José Luis Martí, Pablo Magaña, Josh Milburn, José Juan Moreso, Serena Olsaretti, Andrew Williams and two anonymous reviewers for their insightful criticism of my argument.

Disclosure Statement

No potential conflict of interest was reported by the author(s).

Funding

This work was supported by the Fundação para a Ciência e a Tecnologia under Grant [SFRH/BPD/110642/2015]; by the Agència de Gestió d'Ajuts Universitaris i de Recerca under Grant [BP 2019 00035]; and by a Visiting Fellowship at the Centre de recherche en éthique, Université de Montréal. Early versions of this article were presented at the 2019 MANCEPT panel 'Just Animals? The Future of the Political Turn in Animal Ethics' and the Université de Montréal's Centre de recherche en éthique. This paper would have been much worse without the comments I received at those events.

References

Berlin, I. (1969/1990). *Four essays on liberty*. Oxford University Press.

Birch, J., Burn, C., Schnell, A., Browning, H., & Crump, A. (2021). *Review of the evidence of sentience in cephalopod molluscs and decapod crustaceans*. LSE Consulting.

Bramble, B. (2021). Painlessly killing predators. *Journal of Applied Philosophy*, *38*(2), 217–225. https://doi.org/10.1111/japp.12461

Bratman, M. (1997). Responsibility and planning. *The Journal of Ethics*, *1*(1), 27–43. https://doi.org/10.1023/A:1009703818699
Burgis, J. L. (2018). *Making covenants with brute beasts: Making room for non-human animals in a contractualist framework* [Doctoral dissertation]. University of Miami.
Carter, I. (1999). *A measure of freedom*. Oxford University Press.
Cochrane, A. (2009). Do animals have an interest in liberty? *Political Studies*, *57*(3), 660–679. https://doi.org/10.1111/j.1467-9248.2008.00742.x
Cochrane, A. (2012). *Animal rights without liberation*. Columbia University Press.
Cochrane, A. (2013). Cosmozoopolis: The case against group-differentiated animal rights. *Law, Ethics and Philosophy*, (1), 127–141.
Cochrane, A. (2018). *Sentientist politics. A theory of global inter-species justice*. Oxford University Press.
Cowen, T. (2003). Policing nature. *Environmental Ethics*, *25*(2), 169–182. https://doi.org/10.5840/enviroethics200325231
Donaldson, S., & Kymlicka, W. (2011). *Zoopolis. A political theory of animal rights*. Oxford University Press.
Dretske, F. (2006). Minimal rationality. In S. Hurley & M. Nudds (Eds.), *Rational animals?* (pp. 107–116). Oxford University Press.
Dworkin, G. (1970). Acting freely. *Noûs*, *4*(4), 367–383. https://doi.org/10.2307/2214680
Dworkin, R. (2000). *Sovereign virtue: The theory and practice of equality*. Harvard University Press.
Elmore, B. (2021). A critique of scanlon on the scope of morality. *Between the Species*, *24*(1), 145–165.
Faria, C. (2014). Equality, priority, and nonhuman animals. *Dilemata*, (6), 225–236.
Faria, C. (2016). *Animal ethics goes wild: The problem of wild animal suffering and intervention in nature* [PhD dissertation]. Pompeu Fabra University.
Fischer, J. M. (1994). *The metaphysics of free will: An essay on control*. Blackwell.
Fischer, J. M., & Ravizza, M. (1998). *Responsibility and control: A theory of moral responsibility*. University Press.
Frankfurt, H. G. (1971). Freedom of the will and the concept of a person. *The Journal of Philosophy*, *68* (1), 5–20. https://doi.org/10.2307/2024717
Garner, R. (2013). *A theory of justice for animals: Animal rights in a nonideal world*. Oxford University Press.
Gibson, J. J. (1979). The theory of affordances. In J. J. Gibson (Ed.), *The ecological approach to visual perception* (pp. 127–137). Houghton Mifflin.
Giroux, V. (2016). Do animals have an interest in liberty? *Journal of Animal Ethics*, *6*(1), 20–43. https://doi.org/10.5406/janimalethics.6.1.0020
Glock, H.-J. (2009). Can animals act for reasons? *Inquiry*, *52*(3), 232–254. https://doi.org/10.1080/00201740902917127
Goodin, R. E. (1995). *Utilitarianism as a public philosophy*. Cambridge University Press.
Groff, Z., & Ng, Y.-K. (2019). Does suffering dominate enjoyment in the animal kingdom? An update to welfare biology. *Biology & Philosophy*, *34*(4), 40. https://doi.org/10.1007/s10539-019-9692-0
Hare, R. M. (1981). *Moral thinking: Its levels, method and point*. Oxford University Press.
Hobbes, T. (1651/2008). *Leviathan*. Oxford University Press.
Holtug, N. (2007). Equality for animals. In J. Ryberg, T. S. Petersen, & C. Wolf (Eds.), *New waves in applied ethics* (pp. 1–24). Palgrave Macmillan.
Hooker, B. (2000). *Ideal code, real world*. Oxford University Press.
Horta, O. (2010). Debunking the idyllic view of natural processes: Population dynamics and suffering in the wild. *Télos*, (17), 73–88.
Horta, O. (2013). Zoopolis, interventions and the state of nature. *Law, Ethics and Philosophy*, (1), 113–125.
Horta, O. (2016). Egalitarianism and animals. *Between the Species*, *19*(1), 108–144.
Horta, O. (2017). Population dynamics meets animal ethics: The case for aiding animals in nature. In G. Garmendia & A. Woodhall (Eds.), *Ethical and political approaches to nonhuman animal issues: Towards an undivided future* (pp. 365–389). Palgrave Macmillan.
Johannsen, K. (2017). Animal rights and the problem of *r*-strategists. *Ethical Theory and Moral Practice*, *20*(2), 333–345. https://doi.org/10.1007/s10677-016-9774-x

Johannsen, K. (2020). *Wild animal ethics. The moral and political problem of wild animal suffering*. Routledge.

John, T., & Sebo, J. (2020). Consequentialism and nonhuman animals. In D. W. Portmore (Ed.), *Oxford handbook of consequentialism* (pp. 564–591). Oxford University Press.

Korsgaard, C. M. (2018). *Fellow creatures. Our Obligations to the other Animals*. Oxford University Press.

Kramer, M. H. (2003). *The quality of freedom*. Oxford University Press.

Lovett, F. (2010). *A general theory of domination and justice*. Oxford University Press.

Low, P. (2012). *The Cambridge declaration on consciousness*. Retrieved February 7, 2022, from http://fcmconference.org/img/CambridgeDeclarationOnConsciousness.pdf

MacCallum, G. C. (1967). Negative and positive freedom. *The Philosophical Review*, *76*(3), 312. https://doi.org/10.2307/2183622

McMahan, J. (2010). The meat eaters. *New York Times: The Opinionator*, 2021, June 22.

McMahan, J. (2015). The moral problem of predation. In A. Chignell, T. Cuneo, & M. Halteman (Eds.), *Philosophy comes to dinner: Arguments on the ethics of eating* (pp. 268–294). Routledge.

Meijer, E. (2019). *When animals speak. Toward and interspecies democracy*. New York University Press.

Mill, J. S. (1859/2001). *On liberty*. Kitchener.

Moen, O. M. (2016). The ethics of wild animal suffering. *Etikk i Praksis*, *10*(1), 91–104. https://doi.org/10.5324/eip.v10i1.1972

Ng, Y.-K. (1995). Towards welfare biology: Evolutionary economics of animal consciousness and suffering. *Biology & Philosophy*, *10*(3), 255–285. https://doi.org/10.1007/BF00852469

Nozick, R. (1971). *Anarchy, state, and utopia*. Basic Books.

Nussbaum, M. (2006). *Frontiers of justice: Disability, nationality, species membership*. Belknap Press.

Otsuka, M. (2003). *Libertarianism without inequality*. Clarendon Press.

Paez, E. (2019). On the importance of species for rule-consequentialism: A reply to Galvão. *Utilitas*, *31* (2), 179–190. Impact factor (SJR): 0.82 Q1. https://doi.org/10.1017/S0953820818000183

Paez, E. (2020). A Kantian ethics of paradise engineering. *Analysis*, *80*(2), 283–293. Impact factor (SJR): 0.71 Q1. https://doi.org/10.1093/analys/anz077

Paez, E. (2022). A republic for all sentients: Social freedom without free will. *Pacific Philosophical Quarterly*, *103*(3), 620–644. Impact factor (SJR): 1.071 Q1. https://doi.org/10.1111/papq.12351

Palmer, C. A. (2010). *Animal ethics in context*. Columbia University Press.

Pearce, D. (1995/2015). *The hedonistic imperative*. Retrieved February 7, 2022, from https://www.hedweb.com

Pearce, D. (2009). *Reprogramming predators*. Retrieved February 7, 2022, from https://www.hedweb.com/abolitionist-project/reprogramming-predators.html

Pearce, D. (2015). A welfare state for elephants? A case study of compassionate stewardship. *Relations: Beyond Anthropocentrism*, *3*(3.2), 153–164. https://doi.org/10.7358/rela-2015-002-pear

Pettit, P. (1993). *The common mind. An essay on psychology, society and politics*. Oxford University Press.

Pettit, P. (1997). *Republicanism. A theory of freedom and government*. Oxford University Press.

Pettit, P. (2001). *A theory of freedom. From the psychology to the politics of agency*. Polity Press.

Pettit, P. (2015). Freedom: Psychological, ethical, and political. *Critical Review of International Social and Political Philosophy*, *18*(4), 375–389. https://doi.org/10.1080/13698230.2015.1033861

Rawls, J. (1971). *A theory of justice*. Harvard University Press.

Rawls, J. (1993). *Political liberalism*. Columbia University Press.

Raz, J. (1986). *The morality of freedom*. Clarendon Press.

Rowlands, M. (2012). *Can animals be moral?* Oxford University Press.

Rowlands, M. (2019). *Can animals be persons?* Oxford University Press.

Scanlon, T. M. (1998). *What we owe to each other*. Belknap Press.

Sebo, J. (2017). Agency and moral status. *Journal of Moral Philosophy*, *14*(1), 1–22. https://doi.org/10.1163/17455243-46810046

Singer, P. (1979/2011). *Practical ethics* (3rd ed.). Cambridge University Press.

Talbert, M. (2006). Contractualism and our duties to nonhuman animals. *Environmental Ethics*, *28*(2), 201–215. https://doi.org/10.5840/enviroethics200628232

Taylor, C. (1979). What's wrong with negative liberty? In A. Ryan (Ed.), *The idea of freedom: Essays in honour of Isaiah Berlin* (pp. 175–193). Oxford University Press.
Tomasik, B. (2009). *How many wild animals are there?* Retrieved February 7, 2022, from http://reducing-suffering.org/how-many-wild-animals-are-there/
Tomasik, B. (2015). The importance of wild-animal suffering. *Relations: Beyond Anthropocentrism, 3*(2), 133–152. https://doi.org/10.7358/rela-2015-002-toma
Varner, G. E. (2012). *Personhood, ethics, and animal cognition. Situating animals in hare's two-level utilitarianism*. Oxford University Press.
Watson, G. (1975). Free agency. *The Journal of Philosophy, 72*(8), 205–270. https://doi.org/10.2307/2024703
Wilcox, M. G. (2019). Animals and the agency account of moral status. *Philosophical Studies, 177*(7), 1879–1899. https://doi.org/10.1007/s11098-019-01289-x
Wilcox, M. G. (2020). The intrinsic value of liberty for non-human animals. *The Journal of Value Inquiry, 55*(4), 685–703. https://doi.org/10.1007/s10790-020-09762-1
Wolf, S. (1990). *Freedom within reason*. Oxford University Press.
Woodard, C. (2019). *Taking utilitarianism seriously*. Oxford University Press.

Vulnerability and the Ethics of Environmental Enhancement

Catia Faria

ABSTRACT

In this paper, following the taxonomy developed by Mackenzie, Rogers and Dodds of different sources and states of vulnerability, I claim that wild animals are inherently and situationally vulnerable. This is because they can experience suffering as a response to certain internal and external states and have a high exposure to, and a low capacity to cope with, harmful natural processes. From this it follows that we have a moral obligation to support and assist individuals who are occurrently vulnerable and to reduce the risk of dispositional vulnerabilities becoming occurrent in the future, by endorsing some form of what I call 'environmental enhancement'. Finally, I pay critical attention to how to prevent interventions aimed at ameliorating vulnerability from paradoxically generating pathogenic vulnerabilities.

'Vulnerability' is the new buzzword. From discussions in ethics, bioethics and feminist philosophy to everyday moral discourse, 'vulnerability' has become a key-concept in thinking about what we owe to others and how best to comply with such obligations. As expected, 'vulnerability' has also recently emerged in debates in animal ethics. Debates have focused mainly on how the life of nonhuman animals under human control is conditioned by vulnerability (e.g. Beauchamp et al., 2012; Deckha, 2015; Ferdowsian & Choe, 2013; Ferdowsian & Fuentes, 2014; Johnson, 2013; Martin, 2021; Pick, 2018; Satz, 2009; Thierman, 2011), though a few authors have also acknowledged the vulnerable condition of animals in the wild (Martin, 2021), especially in regard to their vulnerability to anthropogenic climate change (C. Palmer, 2021). Despite a certain degree of disagreement, all authors concur with the minimal claim that if there is a moral obligation to protect the vulnerable, then that obligation should extend beyond the human species. I take this significant body of work as providing enough basis for me to assume the minimal claim here. The aim of this paper is thus not to provide an argument for animal vulnerability in general, but rather to assess the viability of 'vulnerability talk' to account for the plight of wild animals in particular. That is, my aim is to assess the capacity of an ethics of vulnerability to account for the specific situation of vulnerability of wild animals and the specific moral obligations that follow from that. The latter, I think, still remains a vastly unexplored issue.

I proceed as follows. First, based on the taxonomy developed by Mackenzie et al. (2014) of different sources and states of vulnerability, I claim that animals living in the wild are inherently and situationally vulnerable (Section 2). Second, I claim that we have a moral obligation to assist those individuals who are occurrently vulnerable and to reduce the risk of dispositional vulnerabilities becoming occurrent in the future, by endorsing some form of what I shall call 'environmental enhancement' (Section 3). Finally, I pay critical attention to how to prevent interventions aimed at ameliorating vulnerability from paradoxically generating pathogenic vulnerabilities (Section 4).

1. Wild Animal Vulnerability

Trichomoniasis is a worldwide parasitic disease commonly found in birds, such as pigeons and doves. While writing this paper, Birdie, a wild young turtle dove was diagnosed with the infection of *Trichomonas gallinae* and, after a few days of struggling with the disease, sadly died. Birdie, their fellow doves and pigeons, as well as other animals with whom we share a sentient condition, are vulnerable beings. They all have bodily and material needs and are exposed to injury, illness and death. In the literature, this is commonly referred to as the universalist (or ontological) conception of vulnerability: the view according to which to be the sort of entity that we are is to be a fragile animal body with an embedded susceptibility to suffering (e.g. Butler, 2016; Fineman, 2008; MacIntyre, 1999; Nussbaum, 2006; Shildrick, 2002; Turner, 2006). Vulnerability in the universalist sense applies to the set of all embodied individuals. On another extended view, known as the contextualist view, the universalist approach, by being trivially true, adds nothing to the moral discussion. If the concept of vulnerability is to play a normative role, it must be able to grasp something particular about a set of individuals that makes them more susceptible to suffering than others and toward whom we have moral obligations distinct from those we have toward ordinary individuals (Goodin, 1985; Hurst, 2008; Levine et al., 2004; Luna, 2009; Macklin, 2003; Schroeder & Gefenas, 2009). Thus, one might ask: is Birdie's condition (and their wild companions') fully captured by a universalist conception of vulnerability or is there instead something specific about their circumstances which distinguishes them from the rest of the 'ordinarily vulnerable'[1]? The goal of this section is to answer this question.

I follow Mackenzie et al. (2014) in thinking that a robust ethics of vulnerability must be able to overcome this alleged tension between different conceptions of vulnerability and to accommodate both universalist and contextual aspects of the vulnerable condition. That is, it must be able to acknowledge our equal susceptibility to harm as well as to flag, without contradiction, the inequalities that render some individuals more susceptible to being harmed than others. The latter will allow us to identify populations with a high risk of harm and hence to prioritize interventions. The former will enable us to recalibrate the normative significance of central concepts within moral theory at large, for reasons that will become apparent in the last section of this paper.

With that in mind, Mackenzie, Rogers and Dodds put forward a helpful taxonomy of different overlapping sources and states of vulnerability. The first source of vulnerability is intrinsic to simply being a human. As human bodies we have certain physiological, affective and social needs and we are dependent on others to meet them. We are 'vulnerable to hunger, thirst, sleep deprivation, physical harm, emotional hostility, social isolation, and so forth' (2014, p. 7). Call this *inherent vulnerability*.

The second source of vulnerability is contextual. That is, it is contingent upon the specific 'personal, social, political, economic or environmental situations of individuals or groups' (2014, p. 7). For instance, though every sentient body is susceptible to suffering caused by a natural catastrophe, the level of exposure to, and the capacity to cope with the consequences of, a natural catastrophe will differ among individuals. Call this *situational vulnerability*.

One might worry that vulnerability so understood might collapse into sentience. Sentience, as defined by the capacity to have mental states with positive or negative valence, is a necessary condition for someone to experience harms and benefits. While 'sentience' encompasses the whole spectrum of valenced experience, 'vulnerability' captures the distinctive state or condition of being susceptible to harm. Thus, sentience is a necessary condition for someone to qualify as vulnerable, both inherently and situationally. Yet, sentience is not a sufficient condition for someone to qualify as situationally vulnerable. Though all sentient individuals are inherently vulnerable, given their specific circumstances, not all of them are situationally vulnerable or some are more situationally vulnerable than others. An appeal to sentience fails to identify who is more liable to being harmed and thus it does not allow us to prioritize interventions.

Of particular importance is a subset of situational vulnerabilities which may originate in botched interventions. In particular, those resulting from interpersonal or socio-political relationships which undermine individuals' autonomy (Mackenzie et al., 2014, p. 9). Call this *pathogenic vulnerability*. Vulnerabilities may take different forms. Occurrent vulnerability refers to a vulnerability that is actual. For instance, the vulnerability possessed by a pregnant person at childbirth. Dispositional vulnerability refers to a vulnerability that while potential, it is not yet a source of harm. For instance, the vulnerability arising from childbirth possessed by all pregnant persons. The normative implication that follows is that we ought to assist 'the occurrently vulnerable and to reduce the risk of dispositional vulnerabilities from becoming occurrent' (2014, p. 8).

Though Mackenzie, Rogers and Dodds's taxonomy is intended to capture the phenomenon of human vulnerability exclusively, it has been applied to the nonhuman case in a fairly straightforward way, particularly in the context of research (Johnson, 2013). Of special interest for our purposes is Clare Palmer's application of the account to examine wild animal vulnerability to climate change (C. Palmer, 2021). According to Palmer, wild animals are, as sentient bodies, inherently vulnerable. They are susceptible to suffer from injury, illness, and interspecific aggression. It is, however, some individuals or groups of individuals' exposure to anthropogenic climate change that makes them particularly vulnerable in the relevant way. Climate change is a source of situational vulnerabilities for wild animals thus generating moral obligations of support and assistance.

I concur with Palmer that insofar as anthropogenic climate change is likely to threaten the health and livelihood of wild animals it may be a source of situational vulnerability. Yet, I believe that her view suggests that in the absence of such a threat, wild animals could accurately be described as falling under the category of the 'ordinary vulnerable'. She says:

> As embodied beings, sentient animals are, like humans, inherently vulnerable to suffering from disease, injury, and predation. Climate change, however, on this account, is a kind of situational vulnerability to which wild animals are exposed to different degrees and to which they have different abilities to respond. (Palmer 2021, p. 4)

In other words, Palmer acknowledges wild animals' vulnerability in the universalist sense, ('ordinarily vulnerable'). However, under her view, only when faced with the threat of climate change do wild animals become vulnerable in the contextualist sense ('more than ordinarily vulnerable'). This downplays wild animals' particularly high exposure to, and extremely low capacity to cope with, harmful natural processes, and it completely ignores human disregard for wild animal suffering. As I will argue below, the explanatory power of Mackenzie, Rogers and Dodds' taxonomy enables us to construe wild animals as both inherently and situationally vulnerable beings, regardless of whether anthropogenic climate change poses a threat to their well-being. If this is right, then our obligations to support and assist wild animals are stronger than ordinarily supposed.

Like humans, wild animals are *inherently* vulnerable. They can experience suffering as a response to certain internal and external states. Due the fragility intrinsic to their sentient condition, presently existing sentient beings are all inherently vulnerable to hunger, thirst, sleep deprivation, physical harm, and often to emotional hostility and social isolation. Now, unlike humans[2], wild animals have a high exposure to natural processes as a result of living in Darwinian ecosystems[3] where suffering and premature death are endemic. Even though human populations, particularly those with limited assets, are still exposed to some natural hazards, human societies have overall increasingly become less vulnerable to the impact of the natural environment. Consider child and infant mortality. While in the human case child and infant mortality have been declining dramatically (Roser et al., 2019), they remain extremely high for animals that live in the wild.

This is largely explained by the reproductive strategy followed by most species and how their growth is affected by various ecological factors. Contrary to humans, the overwhelming majority of wild animals maximize the chances of survival by producing a huge number of offspring. Given the limited amount of resources available in the environment, they have low survival rates, with most individuals dying shortly after birth (e.g. Ng, 1995; Stearns, 1992). On average, each parent is substituted in the next generation only by one individual (Ng, 1995). The rest die, often, painfully. They are torn apart by predators, frozen and starved to death, or, like Birdie, are internally consumed by parasites and disease. Moreover, even the animals that do survive to adulthood live in a constant state of fear and insecurity, as they stressfully prepare for the next catastrophe: they are systematically injured, live in places with a high density of predators and parasites, and have to endure constant variations of food, water and temperature[4].

This suggests that natural processes, as they currently exist, are likely to be net harmful for most wild animals (Faria, 2023; Horta, 2010a, 2017, Johannsen, 2020, Ng, 1995; Soryl et al., 2021; Tomasik, 2009/2015) such that their basic needs are insufficiently, if at all, satisfied. Wild animals are thus *situationally* vulnerable in virtue of their specific environmental circumstances and their high exposure to, and low capacity to cope with, natural processes. This entails that wild animals' situational vulnerability is not reducible to the threat of climate change even though climate change might constitute an additional source of situational vulnerability.

Notice that by 'capacity to cope with natural processes' I mean the individual capacity to become less liable to adverse environmental conditions over the course of a lifetime and not the ability of the species to adapt over large time scales. Focusing on species adaptation neglects the magnitude of adaptive intermediate harms for many generations. The process by which a certain trait prevails is built on a massive reproductive failure, hence, on the suffering and death of all individuals who cannot adapt to their environment, generation after generation. To assess the situational vulnerability of a certain population, either animal or human, what matters is the capacity to cope of currently existing individuals[5].

Crucially, wild animals' situational vulnerability is not only caused by their inability to interfere with natural processes for their own benefit[6], but it is also exacerbated by human disregard for wild animal suffering. While our species has managed to reduce the impact of environmental stressors on human well-being, it has remained unjustifiably oblivious to the fact that they remain a cause of suffering and death to a vast number of wild animals. Wild animals' situational vulnerability thus arises from adverse environmental circumstances in the wild as well as from a particularly unsympathetic moral and political context.

This implies that wild animal vulnerability is both occurrent and dispositional. That is, as they presently exist, wild animal individuals are in dire need of support and assistance and thus in a state of actual vulnerability. They are *occurrently* vulnerable. Yet, as Mackenzie, Rogers and Dodds highlight, a certain group 'remains dispositionally vulnerable in societies in which such support is not guaranteed' (2014, pp. 8–9). Given that present societies are typically unconcerned with nonanthropogenic wild animal suffering, both extant and future wild animals are potentially vulnerable to persistent natural harms and to human neglect. They are, therefore, *dispositionally* vulnerable.

Future wild animals have a particularly high level of dispositional vulnerability due to the high risk that their current state of affairs will be reproduced in the design of future ecosystems. This is a path already initiated by the so called 'rewilding' approach to ecosystem engineering, according to which the best course of action is to redesign present ecosystems in order to bring them closer to an original state that existed prior to human intervention. Rewilding is the process of reintroducing organisms or restoring ecological processes in ecosystems where they have disappeared or are considered dysfunctional. A paradigmatic example of rewilding is the reintroduction of species, especially predators, into an ecosystem, often resulting in significant negative costs for the lives of many sentient beings (Horta, 2010b; Shelton, 2004).

The virtual absence of moral controversy surrounding the impact of 'rewilding' on wild animal welfare suggests that other, more hypothetical scenarios such as terraforming or directed panspermia would be similarly encouraged in the future, were the technology to become available. To the extent that many terraforming interventions entail spreading sentient life or changing the environmental conditions that might generate sentient life on other planets, they are morally problematic. In a world indifferent to wild animal suffering, it is highly likely that a terraformed planet, by recreating current Darwinian ecosystems, would harbor as much or more suffering than currently exists on earth. If humans decide to embark on such enterprises of ecosystem engineering and spread life to other planets, suffering is likely to multiply on an astronomical scale (Tomasik, 2014 /2018; O'Brien, 2021). This is worrying, since it is not only unlikely that wild animals' occurrent vulnerabilities will fail to be counteracted now, but there is a high risk of wild animals' dispositional vulnerabilities becoming occurrent, and significantly aggravated, in the future.

Acknowledging this does not entail opposing ecosystem engineering *per se*. Ecosystem engineering is, in principle, neutral with respect to the effects on wild animal well-being. This is not to downplay the complex network of power dynamics in which science and technology are typically embedded and its propensity to exploit the worse off. Instead, it is to claim that such knowledge be put at the service of the most vulnerable. Indeed, *if* guided by our strongest moral reasons, ecosystem engineering might become a powerful tool to ameliorate wild animals' vulnerability.

In what follows, I will discuss why we have an obligation to pursue this moral end and what complying with our obligation will amount to in practice.

2. Moral Obligations and Environmental Enhancement

As mentioned, Mackenzie, Rogers & Dodds claim that inherent and situational vulnerability give rise to specific moral obligations to reduce occurrent vulnerabilities and to prevent dispositional vulnerabilities from being realized. If wild animals are inherently and situationally vulnerable, then our reasons to support and assist them are stronger than often presumed.

Now, before going into the specifics of our moral obligations to ameliorate wild animal vulnerability, it is necessary to dispel some doubts. Someone might say that the argument I put forward for addressing wild animal vulnerability is a nonstarter insofar as it relies on Mackenzie, Rogers & Dodds' account which is, in turn, deeply flawed. For instance, Angela Martin (2021) has recently expressed a concern regarding the ability of Mackenzie, Rogers & Dodds' account to explain why vulnerability is morally relevant and requires action on our part. According to Martin, 'insofar as their account grounds moral obligations directly in vulnerability – we have obligations toward others precisely because they are vulnerable – it runs into a circularity problem' (Martin, 2021, p. 4). If vulnerability gives rise to moral obligations, she argues, it is because the vulnerable have a claim to be protected. That claim, however, cannot be based on vulnerability itself. It stems from other moral claims, such as those based on preventing harms or wrongs.

I agree with Martin that vulnerability cannot itself ground moral obligations and that insofar as Mackenzie, Rogers and Dodds are committed to this claim their account is called into question. Yet, it is not the case that Mackenzie, Rogers and Dodds endorse such a claim. Faced with the question of why vulnerability gives rise to moral obligations, they explicitly state:

> Within the literature there are two broad responses to this question. The first response identifies vulnerability itself as the source of moral obligation. The second approach does not ground obligation in vulnerability but rather regards vulnerability as a signal that alerts us to obligations arising from other claims, such as those of harm or need. *Our view is that the second of these views is more plausible* [my emphasis], although we do not present an argument for this claim here. (2014, p. 10)

To be sure, Martin acknowledges this passage in a footnote, yet she takes the absence of an argument for the second approach as a reason to dismiss that approach. I believe Martin's move is unwarranted, though. The absence of a full-fledged argument to support the view that vulnerability flags obligations that derive from independent moral claims might give us a reason to remain agnostic with respect to the account's ability to properly explain the normative significance of vulnerability. Yet, it does not give us a reason to attribute to the account a commitment to the opposite view. Insofar as the view which identifies vulnerability itself as the source of moral obligation is being misattributed to Mackenzie, Rogers and Dodds, Martin's criticism misses the target.

Second, and crucially, it is not the case that Mackenzie, Rogers and Dodds have remained silent on this issue. In a previous paper (Rogers et al., 2012), they sketch out, even if in a scattered way, what their view might look like and, more recently, Catriona Mackenzie (Mackenzie, 2016) has argued without reservation or ambiguity that the normative force of vulnerability is grounded on needs. As she puts it:

> My argument, in brief, is that these twin concepts highlight morally salient features of our humanity that, although largely neglected by moral theorists, are nevertheless central to much of our everyday moral discourse and practice and to our notion of obligations. These features are our finitude, the fragility of our embodiment and agency, our equal susceptibility to suffering and harm and our dependence on others. It is because we are finite, embodied, dependent beings that we are vulnerable and have needs. Further, our shared humanity and our mutual dependence grounds obligations to remediate vulnerability and respond to the needs of others where it is in our power to do so. (Mackenzie, 2016, p. 84)

It now becomes clearer how vulnerability, anchored in need, might give moral agents a reason to act. Notice that even if Mackenzie appeals to our 'shared humanity', surely she does not think that sharing a certain DNA is what grounds our reasons to help others. We all share something morally relevant which is our equal susceptibility to being harmed and our dependence on others for assistance. Clearly, these properties are not exclusive of humans but are rather shared with other sentient animals. Once this is acknowledged, Mackenzie's argument for the normative force of vulnerability might plausibly be reconstructed along the following lines. Our shared *sentience* implies that we are all beings with bodily and material needs, and that we are dependent, to various degrees, on external aid to fulfill them. The unmet needs of the vulnerable, and the agent's capacity to fulfill them, generate an obligation to help obtain whatever it is that they need. Hence, one might infer that the normative force of vulnerability does not arise from vulnerability itself but is rather grounded on the *independent* claim that we should respond to the needs of others whenever it is in our power to do so. Contra Martin, I thus think that Mackenzie, Rogers and Dodd's account of vulnerability still holds.

Someone might object to this reconstruction on other grounds by claiming that the satisfaction of human needs does not normally rely on nonhuman assistance and thus a relationship of mutual dependency simply does not obtain between humans and nonhumans. Since, on Mackenzie's account, mutual dependency is conceived as one of the two bases for moral obligations to aid, our reasons for helping other animals become rather weak or non-existent.

Now, moral obligations to aid as given by 'mutual dependency' should not be understood as some sort of reciprocal bargaining, according to which moral agents have self-regarding reasons to help others to moral agents' own benefit (Nussbaum, 2002). Instead, moral obligations to aid are given by other-regarding reasons. It is the existence of needs and individuals' dependence on external aid to meet them, that constitutes a reason for acting so as to fulfill them. Now becomes clear that the objection assumes a narrow understanding of 'mutual dependency'. When we say that members of a group G are mutually dependent, this relationship is not necessarily symmetrical so that members of G are mutually dependent just in case each and every member of G is dependent on each and every other member and to the same extent. This is something quite uncontroversial when considering human beings, since humans, due to their different capacities and environmental circumstances, clearly differ in their degree of dependency on one another. For instance, A may be dependent on B, while B may be dependent on C and C may be dependent on Z, all to varying degrees. For it to be the case that A and B belong to a group of mutually dependent individuals, A and B need not be dependent on *each other*. It need only be true that A and B are dependent on *someone* and to a certain extent. Accordingly, given that nonhuman animals plausibly depend on human animals to meet their needs and that human animals depend primarily on each other to have their needs met, therefore sentient animals – human and nonhuman – are mutually, though asymmetrically, dependent to varying degrees.

In sum: Mackenzie, Rogers and Dodds' account of vulnerability escapes the charge of circularity identified by Martin. In addition, by putting the focus on needs, instead of interests, their account stresses the universal dimension of dependence. It draws attention to the traditionally devalued dependent nature of *all* sentient individuals, including those living in the wild, thereby providing a theoretical analysis of the often overlooked relational nature of wild animals' moral claims. This is important, since relational accounts in animal ethics are usually unsympathetic to the idea of a general obligation to assist wild animals. In what follows, I will assess two central proposals which are relational in nature, and show how Mackenzie, Rogers and Dodds' account of vulnerability might help us better identify which responses to wild animal suffering might be morally appropriate.

According to a popular view in animal ethics, put forward most notably by Clare Palmer (C. Palmer, 2010, 2012, 2015), we are morally required to assist animals in need *if, and only if*, we have a prior morally relevant entanglement with them, where 'prior morally relevant entanglement' refers to any causal relation between an animal's particular situation of exposure to a harm and past human action[7]. Usually, there is no such morally relevant entanglement between human beings and wild animals. Palmer draws on a well-known analogue for support: the case of parents' special obligations toward their children. Even though all children have similar morally relevant capacities, we only have special obligations to assist our own, since we are responsible for putting them in a situation of vulnerability, by having brought them into existence. Likewise, despite their similar levels of suffering or need, we have only special obligations toward animals we have deliberately made dependent and vulnerable through domestication – those within our 'contact-zone'.

Thus, we should assist domesticated animals, but not those in the wild, not because they are threatened by some harmful event, or due to their unfulfilled needs, but because we are responsible for making them vulnerable to certain threats. This reasoning also provides a more complete explanation of Palmer's view, previously discussed in Section 2, on our obligations to remediate wild animals' vulnerability to anthropogenic climate change. An obligation to help arises *not* because wild animals are vulnerable to climate change per se, but only *because* via human-induced emissions, we have put them in a situation of vulnerability to that threat.

Now, if our reasons to assist other individuals are generated by a causal link between present needs and previous human action, there seems to be no requirement to help distant human beings whose need is due to natural causes. Palmer attempts to avoid this implication by claiming that the relevant connections that hold among humans, and which ground special obligations, are not causal in the sense that they can be traced back to human action but, instead, stand for other kinds of political, economic or familial relationships. However, this second version of her account is also under-inclusive, since many human beings who fail to engage in such relations would be excluded as well. The conclusion is then that either there is no obligation to assist distant human beings in need or there is no basis to exclude wild animals from the scope of these obligations.

Notice that Palmer's claim is not that we ought to refrain from helping animals in need, but instead that we are not usually required to do so. In fact, she thinks that in isolated cases involving an immediate encounter with an individual in need, we should help them as it would constitute a display of virtuous dispositions (C. Palmer, 2010, pp. 148–150). Yet, this is puzzling. If immediate encounters elicit certain dispositions to act, then it seems that such dispositions could be elicited more generally in a larger set of contexts, depending on the extent to which moral agents put themselves in a position to have immediate encounters with individuals in need. This becomes troubling for Palmer's view, particularly considering the implications for her central case of wildebeest drowning during migration, a paradigmatic case of wild animals in need, which does *not*, according to her, generate obligations of assistance. Yet, it would seem that by choosing to place themselves in a position of bystander exposure to the suffering of wildebeests (as many tourists do), and by not helping the wildebeest, moral agents would be failing to display the morally adequate dispositions. Palmer's view thus seems implausible inasmuch the adequacy of moral dispositions should not be made to rely on the agent's choice to place themselves in a position to display the relevant moral dispositions. If at all, the process of moral decision-making should be guided by moral dispositions, and not the other way around.

More to the point, it would be a mistake to believe that morally significant needs only arise within the domain of relations of proximity, either spatial, emotional or causal. One option is to reject relational accounts altogether and ground our reasons to assist others simply on the importance of individual interests. Another, more nuanced option, aligned with the spirit of Mackenzie, Roger and Dodds' analysis, is to acknowledge that even though wild animals' needs cannot be traced back to prior human action nor to emotional or spatial closeness, wild animals' moral claims are still relational in nature. They arise from a morally relevant relationship 'constituted by beneficent ties' (Metz & Miller, 2016) between the needy individual and whoever is in a position to meet their needs. That is, if an individual is in need, then that individual immediately enters into the relevant network of connections with the agent. To clarify:

(i) By being mutually needy and interdependent, sentient beings are engaged in a morally relevant relationship with whoever is in a position to meet their needs;
(ii) That morally relevant relationship grounds an obligation to fulfill sentient beings' needs by whoever is in a position to do so;
(iii) Wild animals are in need;

(iv) Human beings are in a position to meet the needs of wild animals;
(v) Therefore, human beings have an obligation to fulfill the needs of wild animals.

Because we all start with a 'fragile embodiment', all sentient individuals have needs and are dependent on external support to meet those needs (universal vulnerability). As sentient beings, wild animals are engaged in a relationship of dependency with moral agents with respect to the meeting of their needs. No other morally significant relationships have to obtain between the parties for moral obligations of assistance to arise. Due to their high exposure to natural processes and their inability to cope, wild animals have an increased risk of suffering harm arising from unfulfilled needs (situational vulnerability). Our general obligations to assist them thus become particularly strong, calling for immediate action.

Another relational path, followed by Donaldson and Kymlicka (2011), is to affirm the existence of morally relevant relationships with wild animals. Yet, different relationships determine different moral obligations (or duties). Following Swart (2005) they claim:

> Domesticated animals are dependent on their relationship to us for their food and shelter, and so we have duties of 'specific' care towards them, whereas wild animals are dependent on their relationship to their natural environment, and so our duties of care in this context are 'non-specific', and are focused on 'making efforts to maintain their living conditions and their dependent relationship with the environment' (Swart, 2005, p. 258). In each case, we have a positive duty to ensure that animals have what they are dependent on. (Donaldson & Kymlicka, 2011, p. 207)

More clearly:

(i) Both domesticated and wild animals are dependent on X, where X is any means to satisfy their needs;
(ii) Regarding whether we are considering domesticated or wild animals, X stands for different things;
(iii) We have a positive duty to ensure that all animals obtain X;
(iv) For domesticated animals, X stands for some 'relationship to human beings'. Positive *specific* duties are generated;
(v) For wild animals, X stands for some 'relationship with the natural environment'. Positive *nonspecific* duties are generated.
(vi) Thus, we have a positive (nonspecific) duty to respect the dependency of wild animals on their natural environment.

Subsequently, Donaldson and Kymlicka deploy the second part of their argument:

Since

(vii) Wild animal populations are sufficiently competent to successfully engage in self-government – that is, they are sufficiently able to exercise all the necessary functions to thrive without external intervention[8];

Then

(viii) The best way to respect the dependency of wild animals on their natural environment is to establish relations of *sovereignty* with wild animal communities.

I agree with [(i)-(vi)], yet I believe we have compelling reasons to question (vii) the diagnosis of wild animal competence to self-govern, as well as (viii) the adequacy of sovereignty rights as a response to the moral significance of wild animal needs.

As previously discussed, suffering and premature death are the norm in nature. This renders the claim that wild animal populations are sufficiently competent to self-govern extremely implausible. Wild animal populations, as some have claimed, are better described as 'failed states' (Cochrane, 2013; Horta, 2013). 'When basic needs are at stake states have a duty to be impartial, so that if state S is for some reason unable to meet the extensive needs of its own citizens, other states have a duty to step in and fill the gap' (Brock & Miller, 2019). Thus, it is not the case that the best way to respect the dependency of wild animals on their natural environment is to establish relations of sovereignty with wild animal communities. The fulfillment of wild animals' needs does not depend, as Donaldson and Kymlicka think, on the preservation of their natural environments, since the preservation of their natural environments amounts to continuous suffering and death for most animals that come into existence. If we have a duty to ensure that their environment provides the satisfaction of their needs, sovereignty rights are not the solution. On the contrary, the adequate way of complying with the requirement to attend to wild animals' needs is some form of what I call:

> *Environmental enhancement*. That is, the modification of non-sentient natural entities and processes with the aim of increasing the net satisfaction of wild animals' basic needs.

Let me flesh out this definition. First, the set of nonhuman animals potentially affected by environmental enhancement includes all those with the capacity for sentience. This means vertebrate animals – fish, amphibians, reptiles, birds and mammals – cephalopods and decapods, and as more evidence becomes available potentially other invertebrates as well (Birch et al., 2021; Low et al., 2012). Second, environmental enhancement should be distinguished from other, related enhancing interventions, such as nonhuman enhancement. Nonhuman enhancement consists in altering the physical or psychological traits of animals through biomedical or technological means with the aim of promoting their interest in what is necessary to restore or sustain health.[9] Environmental enhancement, for its part, consists in modifying non-sentient natural entities and processes which constrain sentient life in the wild, thereby excluding direct intervention on sentient individuals themselves.

Consider again Trichomoniasis, the parasitic disease previously mentioned. While nonhuman enhancement might recommend turning the whole pigeon population, say via 'gene drive', resistant to the parasite, environmental enhancement would favor eliminating it from the actual ecosystem, via, for instance, vaccination, adequate supply of clean water and other forms of improved sanitation. Notice that even though nonhuman enhancement and environmental enhancement may converge in their effects in terms of overall well-being, they significantly differ in the means by which such effects are brought about. To make this point clearer, consider the ordinary case of a storm falling upon a group of wild animals. Nonhuman enhancement (or disenhancement, in this case) might recommend making nonhuman animals less vulnerable to climate conditions by removing or diminishing their capacity to suffer[10]. Alternatively, environmental enhancement would entail the construction of shelter structures, as well as other mobility assisting mechanisms (e.g. bridges), which *enable* wild animals to find their way into safer places. In other words, while nonhuman enhancement is primarily aimed at eliminating individuals' inherent vulnerability, environmental enhancement is aimed at 'merely' mitigating individuals' situational vulnerability, via the fulfillment of their basic needs.

I will put aside the debate on the necessary and sufficient conditions of a basic need and defer to its intuitive understanding. It seems only plausible that any list of sentient beings' basic needs should include at least enough food, water, and a minimal level of health care. In other words, basic needs are needs that must be met or serious harm will result for deprived individuals (Brock & Reader, 2002).

Notice that while Donaldson and Kymlicka's account remains too optimistic regarding the effect of animals' natural environment on their needs, they are committed as a matter of consistency to make room for the satisfaction of wild animals' needs, via the prescription of *environmental enhancement. If* we are morally required to ensure that nonhuman animals obtain whatever it is that they are dependent on in order to meet their basic needs, *and* animals' natural environment is such that often their basic needs are insufficiently satisfied, or not at all, *then* we are morally required to enhance animals' natural environment in order to ensure the adequate satisfaction of their basic needs[11].

Environmental enhancement is therefore, at least, among the most adequate ways to comply with our obligation to ameliorate wild animal vulnerability: we must provide the necessary assistance to those wild animals who are occurrently vulnerable and reduce the risk of dispositional vulnerabilities becoming occurrent in the future. Recall that wild animals' increased risk of suffering harm arising from unfulfilled needs – their situational vulnerability – is generated both by their adverse environmental circumstances in the wild and by human disregard for their plight. Environmental enhancement interventions directly provide assistance to individuals in an occurrent state of need – by mitigating the context-specific, impairing effects of environmental processes – as well as indirectly reduce the probability of systemic natural harms being persistently neglected. Through the spreading of moral concern for wild animals' needs, it becomes more likely that such support and assistance is guaranteed in the future.

3. Intervention in Nature and Pathogenic Vulnerabilities

Environmental enhancement might be objected to on the basis of a moral concern for wild animals' autonomy and agency. Responding to the needs of individuals makes it likely that we will behave in a paternalistic manner toward those we want to help, thereby undermining their autonomy and agency. For instance, insofar as wild animals have a preference for nonintervention – e.g. they don't like human presence, they prefer not to be in contact with human beings or the process of change would be undesirable to them – interventions on their behalf would be considered paternalistic[12]. Thus, we should oppose environmental enhancement and just let individuals take charge of meeting their needs themselves.

The latter concern is echoed by a particularly troubling subset of situational vulnerabilities identified in Mackenzie, Roger and Dodds' taxonomy, referred to as pathogenic vulnerabilities, which may arise 'when a response intended to ameliorate vulnerability has the paradoxical effect of exacerbating existing vulnerabilities or generating new ones' (2014, p. 9). If vulnerability stands for susceptibility to being harmed, then interventions would make wild animals more vulnerable insofar as they would make wild animals more susceptible to having their autonomy undermined and thus, more susceptible to being harmed in that regard. To the extent that environmental enhancement is likely to jeopardize wild animals' autonomy and agency, it might plausibly bring about pathogenic vulnerabilities.

What about the potential large-scale negative effects of interventions? Aren't they also a source of pathogenic vulnerabilities? Though such interventions would qualify under Mackenzie, Rogers and Dodds' taxonomy as a further source of situational vulnerabilities, they would not necessarily qualify as pathogenic. Pathogenic vulnerabilities are supposed to account for the specific concern of 'responding to vulnerability in ways that avoid stereotyping and paternalism' (Mackenzie et al., 2014, p. 13), rather than for a general concern with overall welfare. That is, the category stresses the importance that interventions be guided by the overall aim of promoting autonomy (Mackenzie et al., 2014, p. 15). Moreover, concern for the counterproductiveness of interventions can typically be addressed as a claim about epistemic limitations (e.g. Donaldson & Kymlicka, 2011; Faria, 2023; Horta, 2017; Johannsen, 2020; McMahan, 2010; Nussbaum, 2006). Since human beings are not, in principle, incapable of obtaining the relevant knowledge, such concern turns out to be indistinguishable from that of an informed interventionist. Notice, however, that even informed interventions, insofar as they might be autonomy undermining, would still remain a potential source of pathogenic vulnerabilities. If that is so, they would fail to constitute an appropriate moral response to wild animal vulnerability. It is my belief though, that this objection can be convincingly met.

First, it is necessary to distinguish two different claims that seem to converge in the objection. One claim is that environmental enhancement may *contingently have* counterproductive effects. For instance, *if* poorly conducted, it will undermine the agency and autonomy of wild animals. The other claim is that environmental enhancement will necessarily jeopardize the agency and autonomy of wild animals. Helping others get what they need always results in an impairment of individuals' autonomy and agency.

One may answer the first claim by stressing that paternalism is not a necessary feature of assistance in general, nor of environmental enhancement in particular. In fact, if moral agents fail to take into account individuals' needs for autonomy and agency when they are essential to the wild animals in question, then moral agents would simply be failing to be responding to those individuals' needs. The probability of this happening is of course proportional to the amount of research we decide to carry out about wild animals' needs and the best technical ways to fulfill them. To that effect, some authors have called for the development of a new scientific discipline – welfare biology – which seeks to study the ecological circumstances that affect the welfare of wild animals (Faria & Horta, 2020; Ng, 1995; Soryl et al., 2021), while others have put forward the need for a multispecies ethnographic research into animals' agency (Blattner et al., 2020). This is clearly no easy task, but at the end of the day, the more we know, the less likely we are to get things wrong. It seems plausible to think that a merger between these two proposals would go a long way toward achieving that end.

The second claim calls for a little more attention. In order to proceed, I will consider three different ways of understanding 'autonomy' and argue that none of them can successfully provide grounds to sustain that claim. In fact, I will argue that, from a more relational approach, protecting wild animals' autonomy is not only compatible with environmental enhancement: it will often require it.

The first thing to note is that wild animals are not, by default, autonomous in the sense of self-sufficient or self-reliant, but are dependent on their natural environment to thrive. Now, for the reasons we saw earlier, their natural environment either does not provide for their basic needs or it does so insufficiently. Hence, the purpose of environmental enhancement is to ensure that wild animals have what they need but are unable to get without assistance. In that sense, environmental enhancement does not make wild animals more dependent. It simply enriches by making the environment they already depend on less hostile and more liveable.

Some might further press this point by saying that even if wild animals are not made more dependent by external aid, environmental enhancement still introduces a new source of control over wild animals' choices. In that sense, intervention in nature would undermine the autonomy of wild animals by undermining their self-government or control over their own choices. Yet, as some have argued, a free choice is such that it entails both immunity to external imposition and an adequate range of options (Paez, forthcoming). In particular, a choice is free only if it is immunized by an institutional shield against the interfering power of others. In that sense, environmental enhancement policies implemented by an institutional system that forced us to consider the interests of wild animals in our deliberations would not constitute an imposition that reduces animals' control over their choices. Moreover, insofar as interventions in nature are aimed at enriching the presently poor and reduced options of wild animals, they increase, rather than reduce, the scope of those choices animals have control over. This is particularly clear with regards to environmental enhancement, where the goal is to furnish present environments with the necessary conditions for individual needs to be met, and thus to broaden individuals' range of options. A concern with freedom and autonomy should thus lead us to endorse human-engineered, supportive environments where wild animals' free choices are truly realized.

Now, Mackenzie, Rogers and Dodds suggest that we reject these dominant conceptions of autonomy, as self-sufficiency or self-governance, and reconfigure the concept of autonomy from a feminist point of view. By identifying relations of interdependence and mutual support as defining features of our lives, which have traditionally been relegated to the realm of the 'feminine', we arrive at an alternative, relational conception of autonomy. As they put it:

> Relational theorists argue that autonomy is a socially constituted capacity, which is developed, sustained, and exercised only with extensive social scaffolding and support. Further, because autonomy is a socially constituted capacity, its development can be impaired and its exercise thwarted by exploitative or oppressive interpersonal relationships, and by repressive or unjust social and political institutions. This approach to autonomy is thus premised on the fact of our inescapable dependency on, and hence, vulnerability to, others. (Rogers et al., 2012, pp. 23–24).

I think that Mackenzie, Rogers and Dodds are right to highlight the often overlooked fact that autonomous agents are socially entangled in the aforementioned way. Yet, I do not think that accepting this should commit us to the more controversial claim that autonomy is a socially constituted capacity. As Jules Holroyd has pointed out 'whilst agents may be relational entities, autonomy may not be' (Holroyd, 2009, p. 323). Or, at least, I would add, not in a constitutive sense. While I do believe that certain relational conditions must be met in order for the necessary capacities for autonomous agency to be fully exercised, I do not think that the capacity for autonomous agency is itself relational, in the sense that it obtains *only* if the agent stands in certain social relations. If that were the case, then in the absence of the relevant social relations, many individuals would not qualify as autonomous. Hence, they would not be susceptible to having their autonomy undermined, even if in possession of the relevant cognitive capacities (Holroyd, 2009, p. 335). This seems highly implausible.

In my view, the crucial aspect of relational autonomy is not 'the social nature of the capacity of autonomy itself' —constitutive conception – but rather the extent to which 'social relationships impede or enhance autonomy'—causal conception— (Mackenzie & Stoljar, 2000, p. 22). The development of autonomy may be impaired and 'its exercise thwarted' *not because* it is a socially constituted capacity, but rather because of the extent to which its full realization *contingently* depends on a particularly supportive social network. Applied to our case, the shift in focus from a constitutive to a causal approach to relational autonomy allows us to highlight how wild animals' autonomy is, as a matter of fact, shaped by external social conditions, in particular, how it is dependent on human support for an adequate range of options, without thereby committing us to the rather implausible view that wild animals' autonomous agency could only be attained *through* such support.

Some might say that this causal approach to autonomy is only tangentially relational, offering little new with respect to a more individualistic understanding of autonomy. I disagree. Relational autonomy, even if only causal, casts doubt on the assumption of the self-sufficient, independent self on which the traditional model of autonomy relies. By emphasizing the social structures that contribute to individuals' development, it explains how supporting the needs of others does not necessarily amount to an undermining of individuals' autonomy and agency, and that it may often be a pre-condition for the full exercise of such capacities.

Appealing to a causal relational conception of autonomy within the context of wild animal vulnerability is thus important insofar as it calls into question the widespread misrepresentation of wild animals as distinctively self-sufficient and self-governed individuals, thriving independently of human assistance. Given that wild animals' autonomy-relevant capacities are impaired by external factors – both by adverse environmental factors and extreme human neglect – wild animals are currently lacking in self-governance. Thus, interventions on their behalf cannot qualify as paternalistic.

In particular, insofar as environmental enhancement is sensitive to the fact that certain social supports must be in place in order for the necessary capacities for autonomous agency to be exercised, it qualifies as an 'agency-responsive' (Rogers et al., 2012, p. 25) intervention. By bringing about the conditions that allow autonomy to develop, environmental enhancement prevents serious harms from falling upon individuals, including serious harm of the sort that undermines wild animals' agency. In that sense, environmental enhancement cannot plausibly constitute a source of pathogenic vulnerabilities and may, instead, be construed as a means to building more resilient 'wild animal communities'.

Even if one remains skeptical about the extent to which other interventions on behalf of wild animals are compatible with fostering wild animals' agency, environmental enhancement is certainly compatible with it. Moreover, an emphasis on the relational elements of autonomy brings to light again the universal (ontological) vulnerability of all sentient beings. Autonomous agents are fundamentally vulnerable beings and, as such, they have needs and are dependent on others to meet them. By recalibrating the significance of central moral concepts, such as autonomy and vulnerability, the vulnerability approach becomes important not only for the purposes of this analysis but also for ethical theory more generally. In particular, it lays the groundwork for a much-needed and potentially fruitful bridge between wild animal ethics and feminist ethics.

Notes

1. The 'more than ordinarily vulnerable' (Sellman, 2005) usually stands for vulnerability in the contextualist sense. I introduce 'ordinarily vulnerable' to refer to the vulnerability as in the universalist sense.
2. Notice that I'm not denying that human beings are situationally vulnerable. My claim is that human beings are *overall* less situationally vulnerable than wild animals.
3. By 'Darwinian ecosystems' I mean ecosystems as they currently exist. That is, ecosystems as they have evolved through the forces of natural selection, with no purpose other than pursuing the reproductive success of biological populations.
4. For an overview of the data see, for example, Faria (2023, pp. 60–85), Johannsen (2020, pp. 13–15), and Soryl et al. (2021, pp. 7–9).
5. Someone might say that if so, individual humans don't cope with natural processes better than other animals, and if they manage to cope, they do so as a species, with the knowledge they acquire over time. Yet, since what matters is the capacity to cope of present individuals, it is true that current wild animals are less capable of coping with natural processes than currently existing humans, since humans have accumulated a knowledge that wild animals presently lack.
6. That is, the inability to interfere with natural processes on the same scale as humans. For instance, to carry out large scale changes such as a redesign of entire ecosystems.
7. The following paragraphs draw from my discussion of Palmer's and Donaldson and Kymlicka's views in Faria (2023).
8. As they put it: 'What sort of competence is needed for sovereignty? We would argue that for wild animals – as indeed for humans – what matters for sovereignty is the ability to respond to the challenges that a community faces and to provide a social context in which its individual members can grow and flourish. And, in this sense, it seems clear that wild animals are competent' (Donaldson & Kymlicka, 2011, p. 175).
9. I put aside the debate about the concept of 'enhancement' and simply follow the definition of Chan and Harris (2011) and Persson and Savulescu (2012).
10. See Johannsen (2020, p. 67).
11. I will leave it open whether we also have an obligation to provide for wild animals' more-than -basic needs. That is, needs the fulfillment of which may be necessary for wild animals to achieve their maximum level of well-being, but whose remaining unmet will not result in serious harm. For instance, the needs for stimulation, amusement, palliative healthcare, as well as to have more than adequate levels of feeding, watering, shelter, bodily temperature and so on. Though I am inclined to think that we do have such an obligation, offering a complete argument to that effect lies outside the scope of this paper.
12. Some might say that, when successful, paternalistically interfering with person A benefits person A despite undermining their autonomy and, therefore, that successful paternalistic interventions do not increase vulnerability. Nevertheless, that presupposes, incorrectly, that

vulnerability relevant harms can only consist in a detraction from well-being. Vulnerability, however, refers as well to susceptibility to other kinds of harms. These include harms consisting in a reduction of autonomy. Therefore, successful paternalistic interventions might make an individual less vulnerable to welfare-related harms and, at the same time, increase their vulnerability to autonomy-related harms.

Disclosure statement

No potential conflict of interest was reported by the author(s).

References

Beauchamp, T., Ferdowsian, H., & Gluck, J. (2012). Where are we in the justification of research involving chimpanzees? *Kennedy Institute of Ethics Journal*, *22*(3), 211–242. https://doi.org/10.1353/ken.2012.0012

Birch, J., Burn, C., Schnell, A., Browning, H., & Crump, A. (2021). *Review of the Evidence of sentience in cephalopod molluscs and decapod crustaceans*. LSE Consulting.

Blattner, C. E., Donaldson, S., & Wilcox, R. (2020). Animal agency in community: A political multispecies ethnography of VINE sanctuary. *Politics and Animals*, *6*, 1–22.

Brock, G., & Miller, D. (2019). Needs in moral and political philosophy. In E. N. Zalta (Ed.), The Stanford encyclopedia of philosophy (Summer 2019 Ed.). https://plato.stanford.edu/archives/sum2019/entries/needs/.

Brock, G., & Reader, S. (2002). Needs-centered ethical theory. *The Journal of Value Inquiry*, *36*, 425–434. https://doi.org/10.1023/A:1021910832459

Butler, J. (2016). Rethinking vulnerability and resistance. In *Vulnerability in resistance* (pp. 12–27). Duke University Press. https://doi.org/10.1515/9780822373490-004

Chan, S., & Harris, J. (2011). Moral enhancement and pro-social behaviour. *Journal of Medical Ethics*, *37*(3), 130–131. https://doi.org/10.1136/jme.2010.041434

Cochrane, A. (2013). Cosmozoopolis: The case against group-differentiated animal rights. *Law, Ethics and Philosophy*, *1*, 127–141. https://raco.cat/index.php/LEAP/article/view/294785

Deckha, M. (2015). Vulnerability, equality, and animals. *Canadian Journal of Women and the Law*, *27*(1), 47–70. https://doi.org/10.3138/cjwl.27.1.47

Donaldson, S., & Kymlicka, W. (2011). *Zoopolis: A political theory of animal rights*. Oxford University Press.

Faria, C. (2023). *Animal ethics in the wild*. Cambridge University Press.

Faria, C., & Horta, O. (2020). Welfare Biology. In B. Fischer (Ed.), *The Routledge handbook of animal ethics* (pp. 455–466). Routledge.

Ferdowsian, H., & Choe, C. (2013). Extending human research protections to non-human animals. In R. Corbey & A. Lanjouw (Eds.), *The politics of species: Reshaping our relationships with other animals* (pp. 232–240). Cambridge University Press.

Ferdowsian, H., & Fuentes, A. (2014). Harms and deprivation of benefits for nonhuman primates in research. *Theoretical Medicine and Bioethics*, *35*(2), 143–156. https://doi.org/10.1007/s11017-014-9288-2

Fineman, M. A. (2008). The vulnerable subject: Anchoring equality in the human condition. *Yale Journal of Law & Feminism*, *20*(1), 1–24.

Goodin, R. E. (1985). *Protecting the vulnerable: A reanalysis of our social responsibilities*. University of Chicago Press.

Holroyd, J. (2009). Relational autonomy and paternalistic interventions. *Res Publica*, *15*(4), 321–336. https://doi.org/10.1007/s11158-009-9090-6

Horta, O. (2010a). Debunking the idyllic view of natural processes: Population dynamics and suffering in the wild. *Télos*, *17*(1), 73–88. https://dialnet.unirioja.es/revista/2311/V/17

Horta, O. (2010b). The ethics of the ecology of fear against the nonspeciesist paradigm: A shift in the aims of intervention in nature. *Between the Species, 13*(10), 163–187. https://doi.org/10.15368/bts.2010v13n10.10

Horta, O. (2013). Zoopolis, interventions and the state of nature. *Law, Ethics and Philosophy, 1*, 113–125. https://raco.cat/index.php/LEAP/article/view/294784

Horta, O. (2017). Animal suffering in nature: The case for intervention. *Environmental Ethics, 39*(3), 261–279. https://doi.org/10.5840/enviroethics201739320

Hurst, S. A. (2008). Vulnerability in research and health care: Describing the elephant in the room? *Bioethics, 22*(4), 191–202. https://doi.org/10.1111/j.1467-8519.2008.00631.x

Johannsen, K. (2020). *Wild animal ethics: The moral and political problem of wild animal suffering*. Routledge.

Johnson, J. (2013). Vulnerable subjects? The case of nonhuman animals in experimentation. *Journal of Bioethical Inquiry, 10*(4), 497–504. https://doi.org/10.1007/s11673-013-9473-4

Levine, C., Faden, R., Grady, C., Hammerschmidt, D., Eckenwiler, L., & Sugarman, J. (2004). The limitations of 'vulnerability' as a protection for human research participants. *The American Journal of Bioethics, 4*(3), 44–49. https://doi.org/10.1080/15265160490497083

Low, P., Panksepp, J., Reiss, D., Edelman, D., & Van Swinderen, B. (2012, July 7). The Cambridge declaration on consciousness. In *Presented at the Francis Crick Memorial Conference on Consciousness in Human and Non-Human Animals*. Churchill College, University of Cambridge.

Luna, F. (2009). Elucidating the concept of vulnerability: Layers not labels. *International Journal of Feminist Approaches to Bioethics, 2*(1), 121–139. https://doi.org/10.3138/ijfab.2.1.121

MacIntyre, A. (1999). *Dependent rational animals. Why human beings need the virtues*. Open Court.

Mackenzie, C. (2016). Vulnerability, needs and moral obligation. In C. Straehle (Ed.), *Vulnerability, autonomy, and applied ethics* (pp. 89–106). Routledge.

Mackenzie, C., Rogers, W., & Dodds, S. (Eds.). (2014). *Vulnerability: New essays in ethics and feminist philosophy*. Oxford University Press.

Mackenzie, C., & Stoljar, N. (Eds.). (2000). *Relational autonomy: Feminist perspectives on autonomy, agency, and the social self*. Oxford University Press.

Macklin, R. (2003). Bioethics, vulnerability and protection. *Bioethics, 17*(5–6), 472–486. https://doi.org/10.1111/1467-8519.00362

Martin, A. K. (2021). Animal vulnerability and its ethical implications: An exploration. *Journal of Applied Philosophy, 38*(2), 196–216. https://doi.org/10.1111/japp.12390

McMahan, J. (2010, September 19). The meat eaters. *The New York Times*.

Metz, T., & Miller, S. C. (2016). Relational ethics. *The International Encyclopedia of Ethics*, 1–10. https://doi.org/10.1002/9781444367072.wbiee826

Ng, Y. K. (1995). Towards welfare biology: Evolutionary economics of animal consciousness and suffering. *Biology & Philosophy, 10*(3), 255–285. https://doi.org/10.1007/BF00852469

Nussbaum, M. C. (2002, December 18). The enduring significance of John Rawls. *Evatt*. https://www.evatt.org.au/post/enduring-significance-john-rawls

Nussbaum, M. C. (2006). *Frontiers of justice: Disability, nationality, species membership*. Belknap Press.

O'Brien, G. D. (2021). Directed panspermia, wild animal suffering, and the ethics of world-Creation. *Journal of Applied Philosophy* [early online view], *39*(1), 87–102. https://doi.org/10.1111/japp.12538

Paez, E. (forthcoming). Wild animal ethics: A freedom-based approach. *Ethics, Policy & Environment*. https://doi.org/10.1080/21550085.2023.2200728

Palmer, C. (2010). *Animal ethics in context*. Columbia University Press.

Palmer, C. (2012). What (if anything) do we owe wild animals? *Between the Species, 16*(1), 4. https://doi.org/10.15368/bts.2013v16n1.4

Palmer, C. (2015). Against the view that we are normally required to assist wild animals. *Relations: Beyond Anthropocentrism, 3*(2), 203–210. https://doi.org/10.7358/rela-2015-002-palm

Palmer, C. (2021). Assisting wild animals vulnerable to climate change: Why ethical strategies diverge. *Journal of Applied Philosophy, 38*(2), 179–195. https://doi.org/10.1111/japp.12358

Persson, I., & Savulescu, J. (2012). *Unfit for the future: The need for moral enhancement*. OUP Oxford.

Pick, A. (2018). Vulnerability. In L. Gruen (Ed.), *Critical terms for animal studies*. (pp. 410–424). Chicago University Press.

Rogers, W., Mackenzie, C., & Dodds, S. (2012). Why bioethics needs a concept of vulnerability. *International Journal of Feminist Approaches to Bioethics*, *5*(2), 11–38. https://doi.org/10.3138/ijfab.5.2.11

Roser, M., Ritchie, H., & Dadonaite, B. (2019, November 19). Child and infant mortality. *Our World in Data*. https://ourworldindata.org/child-mortality

Satz, A. B. (2009). Animals as vulnerable subjects: Beyond interest-convergence, hierarchy, and property. *Animal Law*, *16*(2), 65–122. https://doi.org/10.5565/rev/da.206

Schroeder, D., & Gefenas, E. (2009). Vulnerability: Too vague and too broad? *Cambridge Quarterly of Healthcare Ethics*, *18*(2), 113–121. https://doi.org/10.1017/S0963180109090203

Sellman, D. (2005). Towards an understanding of nursing as a response to human vulnerability. *Nursing Philosophy*, *6*(1), 2–10. https://doi.org/10.1111/j.1466-769X.2004.00202.x

Shelton, J. A. (2004). Killing animals that don't fit in: Moral dimensions of habitat restoration. *Between the Species*, *13*(4), 3. https://doi.org/10.15368/bts.2004v13n4.3

Shildrick, M. (2002). *Embodying the monster: Encounters with the vulnerable self*. Sage.

Soryl, A. A., Moore, A. J., Seddon, P. J., & King, M. R. (2021). The case for welfare biology. *Journal of Agricultural & Environmental Ethics*, *34*(2), 1–25. https://doi.org/10.1007/s10806-021-09855-2

Stearns, S. C. (1992). *The evolution of life histories*. Oxford University Press.

Swart, J. (2005). Care for the wild: An integrative view on wild and domesticated animals. *Environmental Values*, *1*(4), 251–263. https://doi.org/10.3197/0963271054084902

Thierman, S. (2011). The vulnerability of other animals. *Journal of Critical Animal Studies*, *9*(1–2), 182–208. http://www.criticalanimalstudies.org/wp-content/uploads/2009/09/10.-Thierman-S-2011-Issue-1-2-The-Vulnerability-pp-182-208.pdf

Tomasik, B. (2015 [2009]). The importance of wild-animal suffering. *Relations: Beyond Anthropocentrism*, *3*(2), 133–152. https://doi.org/10.7358/rela-2015-002-toma

Tomasik, B. (2014/2018). Will space colonization multiply wild-animal suffering? *Essays on Reducing Suffering*. https://reducing-suffering.org/will-space-colonization-multiply-wild-animal-suffering/

Turner, B. S. (2006). *Vulnerability and human rights*. Pennsylvania State University Press.

Solidarity with Wild Animals

Mara-Daria Cojocaru and Alasdair Cochrane

ABSTRACT

'Solidarity' is a key concept in political movements and usually bears on matters of labour, health and social justice. As such, it is essential in the reproduction and transformation of communities that support their members and protect their interests. It is sometimes overlooked that interspecies solidarity already pertains with a number of domesticated animals, and that people are willing to carry substantial emotional, financial and social burdens to benefit them. There has been even more reluctance to acknowledge wild animals as fellow members of our political communities with whom we might stand in solidarity. The purpose of this paper is to propose that solidaristic relations with certain wild animals are not only possible, but also already exist in some contexts. It explores three different understandings of solidarity, and in probing to what extent they can be applied to wild animals, it identifies some important different roles they can play in 'animal politics', and shows how they can help us reconceive our communities in less anthropocentric terms.

Several scholars have identified a 'political turn' in animal ethics (Cochrane, 2018; Donaldson & Kymlicka, 2011; Garner & O'Sullivan, 2016; Garner, 2013; Milligan, 2015; Palmer, 2010; Smith, 2012; Wissenburg & Schlosberg, 2014). One of the key ideas behind this is that some non-human animals (henceforth 'animals') are not simply creatures with moral status, but fellow *members* of our 'multi-species communities' (see also Midgley, 1983). In part, this is a *descriptive* claim: biologically modern humans have always had socially significant relations with non-human animals. The social significance can stem from direct contact and the various roles non-human animals have played for humans (e.g. as collaborators in hunting, guardians, etc.). It can also stem from the symbolic meaning humans project onto certain animals (e.g. in animistic cultures, or even in contemporary practices such as animal-assisted therapy). But the claim about animal membership is also *normative*: we come to a richer understanding of our obligations toward certain animals if we see them not as passive recipients of our moral agency, but as part of collective projects in pursuit of the common good. Acknowledging their political as well as moral status in this way means that our duties to these animals are not exhausted by prohibitions on killing them or causing them to suffer; instead, we must also consider the forms of *assistance* we can provide to help them flourish, as well as the *collective efforts* we can make in pursuit of goals that extend beyond human ends (Donaldson & Kymlicka, 2011).

Reconceiving our political communities as 'multi-species' and less anthropocentric raises some important challenges. Not least, is the issue of whether familiar normative concepts and political institutions can or should help in building these new communities. In the ongoing research efforts on this front, there has been ample debate about concepts such as 'justice' and 'political representation' (Cochrane, 2018; Garner, 2013; Nussbaum, 2006; Vink, 2020). Less explored is the concept of 'solidarity'. Yet solidarity is key in political communities, and in different ways. Solidarity entails fellow feeling and, in so doing, can serve as a complement to justice (Habermas, 1986). However, solidarity is also more than that: as a unity that mediates between the individual and the community (Scholz, 2008, p. 5) it takes us beyond sympathy and charity, and results in robust commitments (for example, to aid the vulnerable or make sacrifices for others) and practices (like health care or advocacy). These are essential in the reproduction and transformation of the social, in that members of a community can both rely on these commitments and practices for support and adjust them to suit new needs. Solidarity, then, is about communal identification with the other through understanding the perspective of the other. While solidarity comes in different forms, it always entails positive obligations to act and to carry costs for others (Prainsack & Buyx, 2012, p. 346; Rippe, 1998, p. 356; Scholz, 2008, p. 11). Without solidarity, individuals will find it difficult to make the necessary commitments to positive, meaningful action and the sacrifices that are required of co-membership (Cochrane, 2018, pp. 120–126). Solidarity, we might say, makes co-membership possible and durable over time. If certain animals are to be acknowledged as fellow members of our communities, then, it is vital that we stand in solidaristic relations with them. But with which animals should that be?

To date, most of the discussion surrounding animals' membership has focused on *domesticated* animals (Donaldson & Kymlicka, 2011), and when a city like Nottingham organizes pensions for its canine police work (BBC News, 2013; Coulter, 2016, p. 77), we can call this solidaristic without stretching the concept. Further, both the animal charity culture and the pet industry indicate – for better or worse – that private individuals, too, are willing to carry substantial emotional, financial and social burdens to benefit certain domesticated animals. However, there has been more reluctance to see *wild* animals as fellow members of our political communities (Donaldson & Kymlicka, 2011; Regan, 2004).[1] This is likely due to the fact that it is less obvious how solidarity obtains between us and them. For one, unlike domesticated animals who live and sometimes work with us, wild animals often live at a distance from us. Furthermore, while domesticated animals have been selectively bred for certain traits that allow them to live within human settlements, wild animals are usually more lacking in capacities like docility. Both of these facts make it harder to acknowledge the social and solidaristic relations humans have with wild animals.

Yet not all wild animals live at a distance from us or lack sociability. Indeed, the complications of definitively categorizing animals as 'wild', 'domesticated', or something else, are revealed by animals with tangled histories and relationships with humans such as dingoes, pigeons, stray dogs, stray cats, and the many other 'wild' animals (squirrels, rats, badgers, foxes, etc.) who inhabit urban environments. To help us navigate these complications, then, we follow Clare Palmer (2012, 2016, p. 239) in identifying four different forms of wildness, several of which may be relevant for particular wild animals. The 'locational' sense of wildness refers to those animals living (or left to live) outside of our houses, villages and cities (such as ocean-inhabiting whales). The 'dispositional' sense of wildness refers to animals who are not disposed to be 'tame' (lacking in fear and aggression) toward humans (such as crocodiles held captive in zoos). The 'self-willed' understanding of wildness refers to being (relatively) free from human control (such as feral dogs). And the 'constitutive' sense of the term refers to those animals who have not been domesticated (such as garden birds). Our understanding of 'wild animal' thus refers to an animal who exhibits any one or more of these forms of wildness. In going forward, these understandings will be useful to make more precise the ways in which solidarity between humans and wild animals can pertain, and to question some of the socially contingent distinctions between domesticated and wild animals.

The main purpose of this paper is to propose that solidaristic relations with certain of these wild animals are not only possible, but actually already exist in some contexts. This must make us reappraise our view of the multi-species communities we inhabit and should inhabit. And while the possibility and existence of these solidaristic relations does not mean that all animals must automatically be acknowledged as equal fellow members of our communities (perhaps there are countervailing reasons to deny them such status), it does offer an important stepping stone in helping to realize less anthropocentric multi-species communities. Crucially, we argue that solidaristic thinking invites this reappraisal in three ways: by transforming our understanding of the social that undergirds any political community through acknowledgment of our communal identification with certain wild animals; by recognizing the oppressive nature of many of our relations with wild animals, and the necessity of collectively tackling them; and through recognition of the possibility – and importance – of building shared institutions which support all members of the community, no matter their species.

The paper proceeds via four substantive sections. The first section returns to literature in the political turn in animal ethics and argues that the concept of solidarity ought to play a more prominent role in its formulations. The subsequent three sections explore whether solidarity with animals – and with wild animals in particular – is even possible. To do so, three different understandings of solidarity identified by Sally Scholz (2008) – social, political, and civic – are addressed in turn and, in probing to what extent they can be applied to other animals, we also identify important different roles that solidarities can play in a renewed 'animal politics'.

1. Why Solidarity Matters

As referred to above, there has recently been a so-called 'political turn' in animal ethics. Traditionally, animal ethics has been concerned with the 'moral status' of animals, and our duties toward them. For instance, Peter Singer (1995) famously argued that all sentient animals (those with a capacity to experience joy and suffering in their lives) have equal moral status. As such, he argued that we have a duty to consider their interests equally alongside all other sentient beings (including humans), which entails a range of subsidiary duties: to adopt a vegetarian diet; to not buy products which have been tested on them; and so forth. Work in the political turn accepts claims about the moral status of animals and our personal obligations toward them. However, it goes on to suggest that mutually beneficial human-animal relations cannot rely solely on personal transformation: they also require *political* transformation (Cochrane et al., 2018). After all, the vast majority of harms perpetrated against animals have structural (legal, economic, social, institutional and so on) causes, and to meaningfully improve the lives of animals, we must not only change our political systems so as to effectively stop moral atrocities, but truly accommodate animals as members of our shared community or communities.[2]

While this political turn is important and creative, there is a danger that it can take existing political communities for granted and attempt to simply 'stir in' specific animals. The danger of this lies in the anthropocentric understanding of the social that undergirds existing political communities: these are communities that have been created by humans for humans; the identifications they rely on are human-exclusive; the causes they promote are human-exclusive; and the institutions that make them up are human-exclusive both in terms of function and procedures. On top of this, these communities have nearly always been built on the domination of animals: through instrumentalising them for certain human purposes (food, clothing, transport, research etc.); and through inventing certain kinds of animal ('farm animal', 'laboratory animal', etc.) that serve to reinforce these practices. Hence, no animal is likely to receive what they are owed if they are merely stirred into these conceived notions of community: what they are owed will likely be considered optional, of less importance, or simply be blocked by anthropocentric thinking and structures. What the political turn must involve, then, is a *reimagining* of our political communities – pointing the way to *rebuilding* them in less anthropocentric and genuinely multi-species terms.

We claim that solidarity can be instructive here. It points to new ways of envisaging our multi-species communities because it allows for critical discussion and reappraisal of three of their components. First of all, solidarity invites reflection upon the 'social' which must undergird any stable political community. This involves reflection on the communal feelings and identifications which support the community; and draws attention to the contingent nature of the categories that are commonly used to divide and control animals. Secondly, solidarity asks us to think about the appropriate causes we should be pursuing to overturn existing oppressions and rebuild communities in mutually beneficial forms. Furthermore, solidaristic thinking in multi-species communities insists that animals' own perspectives feed into these deliberations. Finally, solidarity also demands consideration of the relevant structures to protect all members. As such, in multi-species communities, we must ask how our institutions are best designed to ensure that the perspectives of non-human members feature prominently.

Crucially, this paper claims that thinking through the presence and possibility of solidarity with *wild animals* is vital to the project of reimagining our political communities. A focus solely on domesticated animals – on those animals who have been selectively bred over generations for traits suited to human purposes – has the danger of taking traditional, extant conceptions of the social for granted and extending political concern solely to those animals not only most like us, but most able to serve us. To rethink mutually beneficial multi-species communities, we need to at least be alive to two possibilities: that they will be made up of and serve animals of various kinds; and that the geographic boundaries of such communities do not neatly overlap with those we have established for contingent reasons (and often by violent means).

But what is solidarity? And is solidarity with animals – and with those commonly seen as wild specifically – even possible? While the concept of solidarity is contested, there are some important shared ideas about it that we will take for granted (see also Coulter, 2016; Scholz, 2008, 2013). For example, it is a key concept in human social and political movements and usually bears on matters of labor, health and social justice. As such, it is importantly different from both charity and disaster relief (Scholz, 2008). Solidarity is also essential in the reproduction and transformation of communities that support their members and protect their interests. It is more than a feeling and less than a political system, but both feelings and institutionalization are components of the full spectrum of solidaristic practices. While it mediates between individuals and communities, it does not aim at establishing homogenous groups; indeed, solidarity does and can prevail in groups whose individuals differ in various ways (e.g. race, sex, class, religion, nationality, etc.). And while it entails positive moral obligations, it does not presuppose or require reciprocity; we can stand in solidarity with individuals, like young infants or individuals who live far from us, who perhaps do not know of our solidarity with them, and/or who are unable to do anything for us (Gould, 2007; Rippe, 1998, p. 358). Hence, when thinking about extending solidarity to animals, it is important to emphasize that it does not operate on the condition of strict sameness (Coulter, 2016, p. 150; Plumwood, 2001, p. 200).

To explore the ways in which solidarity might be present or possible with wild animals in subsequent sections, we draw upon three different types of relation to which solidarity can refer and which Sally Scholz calls 'social', 'political', and 'civic solidarity' respectively (Scholz, 2008, chap. 1).

2. Social Solidarity

One way in which solidarity manifests is as a result of individuals living with and depending upon one another; it is a consequence of individuals forming a group with which they have a shared sense of community and for which they are prepared to make sacrifices. This is what Scholz refers to as 'social solidarity' and it is primarily used as a *descriptive* term: this type of solidarity may exist within groups whose aims or structures we find abhorrent. Crucially for our purpose of reimagining our political communities, social solidarity, together with its accompanying pro-social attitudes of recognition, gratitude, admiration, shared fates, and so on, must be present for communities (multispecies or not) to be durable.

This form of solidarity can seemingly be applied across the species barrier straightforwardly (Coulter, 2016; Scholz, 2013). Not only do all social animals, not just humans, form communities in such ways, but we also already recognize many animals as part of our communities for whom we are willing to carry costs – the most obvious example being our companion animals. And such feelings also often extend to certain other animals who explicitly work for the public good, such as those who work in the police or the military. These domesticated animals are often regarded as part of the community by the public, as evidenced by medals and memorials in their honor, as well as campaigns to ensure improved protection (Hediger, 2013).

Of course, it might be pointed out that such feelings of community do not extend as far as many would like. They are undoubtedly and unfairly partial, often concentrating on 'cute' creatures, or solely on those who can provide obvious instrumental benefits for humans. Indeed, the notion of the social is still mostly defined in anthropocentric terms with these animals just executing human intentions. Even in these cases, however, some humans will go to great lengths to ensure that the animals' proximate (even if not ultimate) interests are considered. And in certain other cases, some humans devote their lives to care for certain domesticated animals in entirely non-instrumental terms, as in the case of farm animal sanctuaries and rescue organizations for abandoned or abused companion animals. Contradictory as these forms of social solidarity may be, attesting to both the morally best in terms of empathy and worst in terms of instrumental profit seeking in people (Coulter, 2016, p. 140), they strongly suggest that interspecies social solidarity exists – and that provides something for communities to build on and extend should that be desirable.

But perhaps this is too quick. Perhaps social solidarity properly conceived must involve some level of identification and sense of community from *all* members – and perhaps this is not an option for any animal. To be clear, the objection here is not that animals lack the *capacities* to either feel a sense of community with or to positively assist humans. Such a claim is dubious given some animals' behavior: service animals, for example, do behave as if they are part of a community with humans, and they do assist members of that community. Furthermore, recall that reciprocity of feeling or assistance cannot be a *strict* requirement of social solidarity, given the implications of this for our community with young children. Rather, the objection here is that few animals do – and none *should* - feel a positive sense of identification with human communities. For as laudable as it may be for people who work with, say, horses or dogs, to strive to really improve their work lives, the horses and dogs might actually much rather be left alone. Indeed, they may rather not depend on humans in the first place. When thinking about extending solidarity to animals who live with and depend on human communities, then, it seems only right to ask, 'what's in it for the animals?' (Birke, 2009; Coulter, 2016, p. 3). For so-called 'abolitionists' (Francione & Charlton, 2015), domestication and a life living amongst humans are not an option because of the dependency they bring about.

By way of response, it is important to reiterate that social solidarity is primarily a descriptive term referring to actual pro-social attitudes and identification – whether appropriate or not. Certain domesticated animals simply *do* identify with the multi-species communities in which they live, showing affection for other (human and non-human) members, and even making sacrifices for them. Furthermore, while a normative evaluation of the abolitionist claim is unimportant to the simple identification of social solidarity, it should be noted that authors in the political turn are keen to explore the possibility of human-animal relations that are mutually beneficial (Donaldson & Kymlicka, 2011). We consider this to be a promising approach, and even one that might allow humans to potentially make up for past injustices.

Yet what about wild animals? When it comes to those who are *locationally* wild, one might expect that forms of social solidarity would be lacking – both in terms of identification and the emotional, financial and other investment associated with it. Can the social really be expanded to those who live far away from us, and thus with whom we have no intimate relations? And these concerns are also present when it comes to animals who are neither tame nor docile, and so *dispositionally* wild. Doesn't social identification require some kind of sociability? It is also unclear why humans should particularly identify with or care about animals who are *constitutively* wild and just happen to pass through our dwellings, or even cause damage to them. More identification and pro-social attitudes could be expected with regard to some of those creatures whose wildness is *self-willed*; after all, they usually still seek human company, at least partly depend on human communities, and can play important social roles. For example, think of the significance of pigeons in many communities who are often welcomed and fed (Amir, 2021), or of those stray dogs (so-called 'riot dogs') who have joined together with activists in political protests in Greece, Chile and elsewhere (BBC News, 2011; Isfahani-Hammond, 2020). On this basis, we will take for granted the presence and possibility of social solidarity with such wild animals, and concentrate on the other three types of wildness here.

Interestingly, identification with animals living (or left to live) wild both in the locational (e.g. ocean dwelling whales) and in the constitutive sense (e.g. garden birds) is actually widespread. It is true that for some, the ways in which they approach these animals are primarily motivated by a certain exoticism stemming from the idea that wild animals are the quintessential 'Other'. Nonetheless, there are also those indigenous communities who possess a very strong sense of *kinship* with them – from symbolic and totemic relations, to the many ways in which the lives of human and animals are intertwined through practices like hunting. And such kinship also extends to animals who are wild in the dispositional sense (like bears or snakes). It is worth noting that even sometimes rather weak identification can be built upon and grow: bird-feeding can go far beyond a Sunday morning pastime and inspire people to identify with birds more concretely (think of the enthusiasm for birds during the COVID-19 pandemic (RSPB, 2021)), and even campaign on their behalf (think of opposition to wind energy (Hogan, 2020)).

Moreover, significant emotional and financial investment in wild animals is certainly present within many communities. Just as in the case of abandoned or abused companion animals, there are individuals who sacrifice potentially more lucrative careers and join together to rescue, care for and, if possible, rehabilitate wild animals of all kinds. Often, such work is supported by other members of the community through donations, including sometimes through leaving them legacies in their wills (sometimes at the expense of other humans) (WWF, n.d.). When we add to these examples the public investment that many societies make related to the science and politics that goes into international conservation efforts, it becomes apparent that meaningful social solidarity in the form of genuine communal identification with any type of wild animal can be observed in a number of ways.

And yet, some might still be skeptical and argue that there cannot be genuine communal identification in the absence of repeated interpersonal interactions: in other words, it might be claimed that true social solidarity requires close proximity, intimacy and sociability, things which many wild animals do not seem to seek with humans.[3] By way of response, it is worth stating three things. In the first place, it is not the case that *all* wild animals live far from us and lack the ability to engage with us socially. As we have seen, the category of wild is fuzzy, heterogeneous and contingent; as such, it necessarily includes many animals who live alongside us (e.g. foxes, badgers, bats, rats, pigeons, migratory birds, etc.), some of whom *do* sometimes seek human company or support (e.g. stray populations of dogs, cats, etc.). Second, and more importantly, we should resist the idea that close proximity, intimacy and sociability are in fact required for communal identification and social solidarity to pertain. After all, the paradigm political communities of our time – states – are enormous and complex groups in which close proximity and intimate social relations between all members is simply impossible. Nevertheless, it makes sense to say that social solidarity exists within these communities on the basis that individuals identify with other members as part of the same community for whom they are willing to make certain sacrifices. Such identification might derive from an understanding of the contributions each member makes to the community, a sense of shared history, shared goals, or something else. But as we have explained, all of these identifications – and their sources – are possible and present with certain wild animals. Finally, none of this should be taken to mean that we are arguing for each and every wild animal to be brought in under the control of 'human' communities. The purpose of thinking through social solidarity with wild animals is not a claim about the expansion of *existing* communities. Instead, it is about reimagining and redrawing our shared political communities in genuinely multi-species terms: by recognizing social solidarity where it exists and might exist, we can undergird our multi-species communities in less anthropocentric terms.

This call for transformation also helps to explain why we should not regard existing solidaristic relations with wild animals as normatively unproblematic. To reiterate, Scholz identifies social solidarity as a descriptive term, and we have been using it to draw attention to the many socially significant relations between humans and wild animals that already exist today. Importantly, the picture that emerges of the concomitant social solidarity is extremely varied and full of tensions; indeed, we should hope for communities to take a more systematic and normatively justifiable approach to these relations. For while social solidarity might be used to describe existing relations, that obviously does not prevent us from attempting to shape those relations in ways that are normatively desirable. We claim that solidarity's focus on work, health and social justice, can help to critically question extant social solidarities and forge improved ones. Is our acknowledgment of animal work too focused on the services provided by certain domesticated animals, and not enough on the work wild animals provide via ecosystem services (e.g. scavenging and population control)? In what ways is human health and food security dependent on the health of wild animals and their habitats (e.g. pollinating crops and aerating soil)? When considering the just distribution of resources within a society, how much weight ought to be given to the claims (e.g. to habitat, territory and even property) of the wild animals who reside there? Tackling these questions will surely be crucial in building and maintaining less anthropocentric and more plausible visions of our multi-species communities.

None of this is to claim that tackling these questions will lead to easy answers; without doubt, the pictures of social solidarity with wild animals that emerge will be contested. For example, at least sometimes, individuals may identify with groups of animals whose interests are not aligned. Are you for the seal pup or for the orca? Are you for the hatchling crow or for the squirrel? How, if at all, can such identifications be reconciled? If we cannot identify with all wild animals, then how should the lines be drawn? There is no doubt that our pro-social attitudes are presently volatile, and too volatile to resolve these issues easily. But in our view, tackling these issues head on – and tackling them through the lens of interspecies solidarity and a transformed and expanded, non-anthropocentric view of community – can be very instructive here.

3. Political Solidarity

The second type of solidarity identified by Scholz (2008) is 'political solidarity'. This form of solidarity refers to when activists unite for a particular political cause or project. In particular, it relates to when activists generate a shared commitment to fight oppression that is human in origin and to develop strategies for success. It is important to see that, here, the normative commitment is prior to the community (of activists); indeed, the former brings the latter into being and can dissolve once the goal has been reached, no long-term identification is needed. Political solidarity can also act as a vehicle for reflecting on what types of communal identification with animals, if any, are appropriate, thus addressing some of the challenges referred to in the previous section with regards to normatively problematic extant forms of social solidarity. Finally, articulating the need for political solidarity with and on behalf of wild animals helps to identify and overcome existing forms of oppression of them, which, in our view, are as vital as the familiar causes around which most animal activists currently unite.

Given that political solidarity relates to a cause to fight *oppression*, it is important to have a clear understanding of what oppression is. Here we follow Scholz's adoption of Iris Marion Young's conception, which sees oppression as consisting 'in systematic institutional processes which prevent some people from learning and using satisfying and expansive skills in socially recognized settings, or institutionalized social processes which inhibit people's ability to play and communicate with others or to express their feelings and perspective on social life in contexts where others can listen' (Young, 1990, p. 38 cited in Scholz, 2008, p. 128). In political solidarity, then, individuals form a 'solidary group' when they respond to such oppression with a commitment to fight it. Individuals can take on various commitments of this kind (Scholz, 2008, p. 142 f.), and such commitments can flow from a variety of sources: conscious choice, emotions, experience, desires and simple circumstances (Scholz, 2008, p. 72). Political solidarity, then, neither relies on adherence to a single particular argument, say about moral standing, nor is it necessarily rational if 'rational' is equivocated with 'in one's self-interest' (Scholz, 2008, pp. 75, 137, 144 f.). After all, political solidarity may require people to make considerable sacrifices. Moreover, Scholz claims (2008, p. 56) that individuals in the solidary group need not suffer from the respective oppression themselves; and, members of the oppressed group do not need to know of the activism or activists, nor be able to develop a reciprocal relation with them.

All of this would suggest that the concept of 'political solidarity' can apply straightforwardly in the context of human-animal relations. Take, as just one really clear-cut example, the vast majority of those domesticated animals used in agriculture. The human institutions and institutionalized processes are such that these animals are prevented from learning the necessary skills to navigate their communities and, a fortiori, from using such skills in satisfying ways. Humans also systematically inhibit most of these animals' ability to play and communicate both with their kin and with humans. And it is fair to say that these animals have almost no chance to express their perspective on social life 'where others can listen'. These kinds of relations are manifestly oppressive and animal activists are right to mobilize politically around them. The same is true for the many ways in which humans impinge on the lives of wild animals: wild animals who suffer from habitat destruction, are held captive in circuses, or who are traded as exotic pets straightforwardly meet the definition of oppression proposed by Young. Furthermore, at a time where human influence is changing living conditions on the whole planet, it is difficult to refuse responsibility for wild animals struggling to adapt to their natural environments.

Yet can we really stand in political solidarity with animals who seem to lack understanding or commitment to the cause for which we are fighting? Scholars of solidarity tend to think that it can be held with groups even when individuals from within that group are not (but could be) committed to the cause (e.g. feminism, see also Scholz, 2008, pp. 123 f., 136; and Rippe, 1998, p. 358). But there still is a question of what happens when those individuals *cannot* be committed to such a cause. Much here hinges on the question of whether animals are capable of the kind of political agency to enable them to share such a political commitment (Coulter, 2016, p. 152; Meijer, 2019; Pepper, 2021; von Essen & Allen, 2020). For some, (Pepper, 2021) animals cannot properly be said to hold such agency given their inability to intentionally affect political institutions, engage in shared intentionality, and imagine alternative political futures. For others (Donaldson & Kymlicka, 2011; Hribal, 2011; Meijer, 2019; von Essen & Allen, 2020), animals can be regarded as holding political agency when we adopt an expanded understanding of the concept which is primarily connected to embodied behaviors, actions and communication. There is clearly not the space to resolve this debate here, but two things are worth pointing out when thinking through its implications for political solidarity. First of all, we might say that in an important sense – namely one that is definitive of their good life and includes their ability to play, communicate and express their feelings – dogs will be dogs, just as cats will be cats. In other words, no attempted change of political consciousness can alter who animals are, and we need not expect the lion to riot with the lamb against human oppression. Nonetheless, skepticism regarding the ability of animals to take up a political cause does not rule out political solidarity *on their behalf*. Indeed, Scholz explicitly argues that while political solidarity *with* animals is not possible, political solidarity *on their behalf* certainly is (Scholz, 2013). Secondly, it seems plausible to adopt a conception of political agency – and hence commitment to a cause – that is not all-or-nothing. For instance, it does not seem outlandish to maintain that an individual can commit to a cause through certain behaviors and practices, such as sabotage, even if that commitment would rightly be regarded as stronger if it was backed up by a reflective rational intention to adopt it. In other words, a scalar understanding of political agency – in which animals (and humans) might have more or less of it – makes some sense.

Ultimately, the worry about animals' agency in political solidarity arises from the fact that humans are always in a position of privilege vis-à-vis animals. As such, animal advocates must question what their activism means for the animals themselves (Birke, 2009; Coulter, 2016); indeed, there is a real risk that a false sense of unity is created, one that obfuscates the animals' real interests. In this way, thinking through politically mobilizing for animals necessarily demands reflection on the kinds of relations (and identifications) we ought, and ought not, to be seeking with animals. Quite obviously, this demands seeking out and consideration of animals' own perspectives and, to the extent that it is possible and helpful, questioning human privilege. Yet Scholz (2008,

p. 158) is also clear that the dangers of privilege should not preclude involvement of the privileged; instead it provides an imperative to make genuine attempts to adopt the perspective of the oppressed group. In so doing, she recalls feminist strategies such as 'world-traveling' (Lugones, 1987). This involves endorsing and willfully exercising a 'flexibility in shifting from mainstream constructions of life [. . .] to other constructions of life [. . .] animated by a [playful] attitude' (Lugones, 1987, p. 3) with the goal of genuine changes in character that lead to the travelers being at ease in these worlds (Lugones, 1987, pp. 14–17). Playfulness makes exploring worlds safe and respectful. It is not about playing certain games, like chess, but approaching shared activities, like navigating a shared environment, with an open attitude that allows these activities to become games. Such games can be transformative (Lugones, 1987, p. 16) and the link with solidarity is that the bonds that are created through such engagements are not those of domination, but of mutual co-creation. If that sounds theoretical and abstract, consider any type of real play between humans and dogs or (particularly young) cows. Not the mechanical or hierarchical type of play (e.g. repeat ball throwing) but that where roles are reversed and the animals are free to actually show themselves. Such interactions could broaden solidarity *on behalf of* animals to be more informed by the animals' own agency and move in the direction of solidarity *with* them, in keeping with the scalar understanding of political solidarity referred to above.

What then of political solidarity with or on behalf of wild animals? Arguably, people who assist individual wild animals who have been injured, or are experiencing hunger or thirst, or actively seeking human help, or even suffering from anthropogenic climate change, can be said to fulfil their commitment to a cause, even if it is one that may be fuzzy at the conceptual outline. Moreover, people who provide sanctuary for dispositionally wild animals rescued from oppressive living conditions in circuses or as pets are often acting out of a political recognition of these wrongs. Sanctuaries that take in abused bears and protect them from harm, also advocate for those animals' right to live without human control, even though they can only do so much for animals once the damage has been done. Indeed, Palmer discusses wild animals as paradigmatic cases for *reparation* precisely because these animals would have lived or would live independently from humans if it wasn't for the harms we have done to them (Palmer, 2010, p. 101). We argue that understanding these oppressive relations as the kinds of cause around which activists ought to politically unite, too, is a vital step in realizing and maintaining more just multispecies communities.

While the actors described above may have political solidarity *on behalf* of these wild animals, is there any possibility of moving toward solidarity *with* them? As you will recall, that would require the animals' own input and perspective on the cause – including through such means as world-traveling. Is that an option with wild animals? On the one hand, social play and role reversal is definitely not the type of interaction we can or should seek with the vast majority of wild animals. But on the other, world-traveling would at the very least require us to be open to reconstruct our human self in the presence of wild animals. And there are important real instances of such 'playfulness' with wild animals. Think, for example, of attempts to condition wild animals in creative ways, and by associative learning, to better navigate shared environments (like using wildlife corridors) or to avoid poachers (like chimpanzees in Sierra Leone, on both examples, see Eberle, 2021). Consider also those 'gamified' activities to encourage humans to join in on collective efforts in the interest of other animals (e.g. awards for counting garden birds, or interactive and fun platforms for analyzing data by citizen scientists, see, for instance, the many projects on Zooniverse). Finally, coming up with sensible and imaginative enrichment strategies for animals in sanctuaries also often exemplifies this kind of playfulness. The goal Lugones envisions for the human sphere, namely, to 'understand what it is to be them and what it is to be ourselves in their eyes' (Lugones, 1987), may of course be additionally complicated in the case of certain animals, but that does not make it impossible for all of them.

If we are to mobilize politically with or on behalf of animals, it is important that we pursue and model the kinds of multi-species communities we hope to facilitate. In other words, it is important that we tackle genuine oppressions in ways that are themselves non-oppressive. And while the injustices of capture, trade, habitat destruction and climate change may be relatively clear-cut, other harms which befall wild animals are more controversial. At the core of many current debates about wild animal suffering is often something different to the kinds of oppressions referred to above. Some thinkers argue that there is something unfortunate about how some species have adapted and evolved in nature (Faria & Paez, 2015; Horta, 2010; Johannsen, 2020). It is pointed out that most wild animals suffer terribly in their lives: the vast majority never even make it beyond infancy due to malnutrition or predation, and those that do are usually struck by prolonged disease, hunger and injury (Horta, 2010 but cf. Browning & Veit, 2021). As a result, some philosophers have proposed that we have a duty to seek ways to reduce such suffering, with various possibilities proposed and evaluated: from preventing some of these sentient animals from coming into existence, to painlessly killing, or altering the genetic make-up of, predators (Bramble, 2021; Moen, 2016). However, since political solidarity is about uniting behind a cause to tackle *oppression* – that is, those particular harms caused by human institutions – it appears unlikely that it could or should coalesce around a duty to tackle this kind of suffering. Of course, this is not to deny that moral agents hold duties to do something about this suffering – perhaps they have certain weaker duties of assistance, for example (Johannsen, 2020). But there does appear good reason to doubt that these are duties grounded in *social justice* and thus the right kind of commitment that follows from political solidarity.

However, maybe such a judgment assumes an already impoverished and anthropocentric understanding of the social. If there are particular wild animals we identify with in our multi-species communities, one can legitimately question why they ought to be left to suffer, even when the harms they endure are 'natural'. Indeed, we might say that their neglect by their communities is a form of oppression in that it comes about both through inhibiting these animals from expressing their feelings in the sense Young admonishes, and through failing to have the appropriate institutional and social processes to mitigate and alleviate their suffering. This is undoubtedly the conclusion we would reach in the human case: if a society were to ignore the harms to a particular section of its population on the basis that they are 'natural', we would justifiably regard that as oppressive. So it is impossible to rule out the validity of mobilizing politically for wild animals even when their suffering is caused by natural processes such as disease, hunger and predation.

Of course, if our political mobilization is both to pursue legitimate causes *and* be non-oppressive, the only way we can answer the question of whether to mobilize for this or that cause is to take heed of animals' own perspectives on these matters. And as miserable as we might imagine them to be, perhaps all, many or some would not wish to be the objects of such activism. This might entail respecting their desire for freedom and nonintervention even in the face of great hardship (Donaldson & Kymlicka, 2011), or it might entail ruling out certain types of intervention, while permitting others. The crucial implication of this discussion, then, is that uniting in political solidarity with and on behalf of wild animals is perfectly possible, but requires us to explore and employ robust and imaginative tools – such as those of world-traveling – in order to understand those animals' perspectives and have these inform our views about the individuals we identify with and the causes we pursue. The next section addresses the importance of institutionalizing processes to that effect within our political structures.

4. Civic Solidarity

The third type of solidarity identified by Scholz is civic solidarity. This is more than a set of pro-social attitudes and also more than a political or social justice movement: it refers to the institutional apparatus to protect vulnerable members within the community and to ensure their continued participation in social life. At the level of nation states and some supranational organizations such as the EU, civic solidarity usually manifests in terms of health services or pension schemes as well as some social services that enable members to access this protection and support. Importantly, these are not institutions of charity, but respond to the justified claims of individuals as members of a political community. Such institutional support can serve as scaffolding for social values and ensures assistance for all and especially the most vulnerable members, even when feelings of charity or resources are low. Importantly, civic solidarity does not just respond to vulnerabilities that stem from institutional oppression; it also deals with the vulnerabilities all community members face whether they stem from age, injury, sickness or any other increased level of dependency. Thinking through civic solidarity with wild animals has important implications for the institutional design of multi-species communities and the processes by which we can take heed of their perspectives within them.

We have already touched on certain health services and pension schemes in the case of service and certain feral, so self-willingly wild, animals above. Public health initiatives like One Health, which explicitly link the health of humans and non-human individuals might be seen as 'embryonic' forms of civic solidarity with all kinds of animal (Rock & Degeling, 2015). It is even possible to see some animal welfare laws – for instance, those that require humans to provide their companion animals with sufficient exercise, or that penalize taking wild animals out of nature without compelling reason – as 'skeletal' forms of civic solidarity, given that they are tied to an institutional framework supported by public funds to prevent certain harms befalling particular animals. Furthermore, recent discussions about animal ombudsmen and -women (e.g. in the city of Lisbon, Brookland, 2019) are promising developments in providing institutionalized representation within our political processes. And finally, conservation is perhaps the most obvious policy area in which political communities already invest considerable resources and money.

However, these embryonic or skeletal forms of civic solidarity are a far cry from the types of institutional arrangement we have created for human members of our societies. Just consider, for instance, the ways in which current societies advocate for animals at the institutional level. Not only is it striking how few dedicated animal advocates exist within our political apparatus, but even when they are present, they do not have the same standing as, say, child protection or immigration case workers. This is because most animal welfare laws are only intended to outlaw the most egregious harms and in ways which ultimately serve human interests. This anthropocentrism is not only explicit in conservation practices, but is also evident in initiatives like One Health, where non-human health is of concern insofar as it serves that of humans. In addition, none of these institutions entails their respective communities making *significant* sacrifices; and none is central to the communities' self-understanding. Instead, the investments they entail are minimal or based on charity. So, while forms of civic solidarity with all animals – whether domesticated or wild – exist, they need to be bolstered hugely if animals' political status is to be made good on. But how is that to be done?

Importantly, establishing robust forms of civic solidarity involves more than just copying moral sentiments onto the institutional canvas; after all, our political structures cannot (and should) not be simple facsimiles of our individual moral beliefs and commitments. To help illustrate this, consider Peter Singer's famous 'child in a pond' thought experiment (1972). By using the example of a near-drowning child most people would no doubt want to rescue, Singer elicits a moral intuition: namely, that we can, by investing relatively little, save the lives of distant others with whom we have no special relations. Some thinkers have argued, quite reasonably, that this idea travels well across the species barrier: if, a sentient being's suffering could be easily prevented, then we have a moral obligation to do just that – whether that being be a puppy drowning in a pond (Moen, 2016), or a frog in a pool. It seems, then, that this establishes collective general duties to minimize all these creatures' respective suffering if it comes at little or acceptable cost to ourselves.

But thinking about wild animal suffering through the lens of civic solidarity invites different responses. For it is clear that the institutional questions we face are very different from those that arise in the drowning child (or puppy or frog) case. First and foremost, solidarity concerns what communities must provide in terms of *social justice*, not beneficence; in other words, what members are owed in terms of social and political rights. Singer, in his example, on the other hand, is concerned with what individuals owe to other individuals. Secondly, unlike in the drowning child case, political communities are not confronted by cases of 'easy rescue'. Communities instead face decisions that are complicated across a number of dimensions. We do not merely confront a single suffering being, but huge numbers of them. Moreover, not only are the reasons behind individuals' predicaments often varied and complicated, but so too are the appropriate remedies, their effectiveness, and the means of funding and distributing them. Sometimes the suffering of wild animals is a result of natural processes, and at others it is the direct result of human policies and institutions; sometimes an intervention will have little effect beyond the individual we are saving, at others it might have all sorts of 'cascade' effects that disrupt the ecosystem in which that individual resides; sometimes an effective remedy is available to us, at others it will require new investment and research; and so on. In summary, the *moral* imperative for one agent to conduct an 'easy rescue' tells us little about how we should design our public services. This is not a question of one individual aiding another, but of a community deciding how to fund, organize and deliver a collective response to the very different needs and claims of the community as a whole.

And the third reason why civic solidarity requires us to think differently about cases of wild animal suffering comes down to the fact that it refers to the institutional framework to protect members of *specific* communities Put simply, a community's institutions cannot protect the basic interests of *everyone*, and must pay special attention to those of its members. While a full account of which wild animals (or any other individuals) warrant membership in which communities is beyond the scope of this paper, a few points are important to make clear. For example, some have suggested that the relations which give rise to special duties of support might be generated from the *benefit* that humans accrue from certain wild animals: for example, think of the benefits to wildlife photographers from 'majestic' animals or to stressed people from garden birds (Jalagania, 2021). And yet, civic institutions exist to serve the *common good* which is more than the aggregate of individual benefits; and importantly, not everything that benefits an individual is beneficial to the public. But of course, and as we have seen already, many wild animals do contribute collective benefits to the public, whether that be through aerating soil, pollinating crops, landscaping, or simply inspiring us to scientific and artistic innovation. Indeed, thinking about our relations with wild animals through the lens of solidarity invites reflection on the real contributions many animals make to the functioning and flourishing of our communities. Such consideration undoubtedly opens the door to a broader and less anthropocentric understanding of community membership. However, there are very good reasons to not make the provision of benefits a necessary condition of membership (think of those members with certain disabilities who nevertheless receive support). But even with this caveat in mind, we can still acknowledge that when benefits are provided, we have good reason to uphold the status of those contributors as fellow members. Crucial in this is that these contributions are accounted for under the perspective of true mutualism, so that both humans and the animals in question have something to gain from the arrangement.

So while thinking about civic solidarity invites tackling wild animal suffering through these different ways (social justice, complexity, membership), what would the institutional frameworks to which it gives rise actually look like? And how would they function? For example, how would they be set up to respond to the aforementioned example of the orca and the seal pup, whose interests very obviously conflict? While there is no hope of detailing fine-grained blueprints of the make-up of these institutional arrangements, three points can nonetheless be made. First, a trans-species welfare state able to rescue each and every human and non-human animal, domesticated and wild in whichever sense, from potential suffering is clearly neither practical nor desirable. It is not *practical* because, even when members are restricted appropriately, it will still be impossible to prevent or alleviate all types of harm for all members. For example, not all seal pups can be saved – but certainly some can. How then should we prioritise? Should, for instance, our institutional priorities rest with those who have been made vulnerable through human policies and actions (Palmer, 2010)? It is also not *desirable* to attempt to eradicate all harms because any such attempts will likely be so intrusive and unwanted that they will backfire. For example, an attempt to save all seal pups from harms may not only have devastating consequences for orcas, and many other species, but may in fact be detrimental to certain seal pups themselves, say by making them too dependent on humans.

As such, and this is the second point, any institutional arrangements that we do create must have input from all members. For example, we need to determine which wild animals count as members in which communities and what their preferences and interests are. Various proposals have been put forward in the 'political turn' literature on how animal agency can be facilitated (Donaldson & Kymlicka, 2011; Meijer, 2019), and/or representation secured (Cochrane, 2018; Vink, 2020) and these need to be utilized and if necessary adapted so that wild animals can also be heard. In addition to these types of political representation, there are also more informal institutions which can play an important role in helping communities be attuned to wild animals' perspectives. One example that might be instructive here is that of wild animal sanctuaries (Abrell, 2021; Pachirat, 2018). Sanctuaries can and should be seen as expressions of solidarity: in the human exclusive case, to provide sanctuary is a political act; and cities who declare themselves sanctuary cities do this out of a recognition of the rights of refugees and at the risk of considerable political costs. In the case of animal sanctuaries, however, people tend to primarily look at things from the perspective of charity and not from social rights. This makes the important work sanctuaries do contingent on feelings of sympathy, and we have already touched on how such feelings can often be conflicted and paradoxical. Instead, then, it would be helpful to reconceive of sanctuaries as sites of learning through world-traveling and, as such, as laboratories of less anthropocentric communities. Such learning can involve humans learning to appreciate mutualistic relations between different species if they get to care for or introduce members of so-called keystone species. But it can also apply to wild animals themselves when sanctuaries work to equip them with the skills needed to return, if possible, to life as locationally wild animals.

Finally, and relatedly, the ultimate point of taking the perspective of wild animals is not so that we know better how to rescue or save individual animals, but so that their interests can properly feed into constructions of the common good. It is impossible to eradicate all suffering, and do away with all risks within any political community, multi-species or not. The aim must be instead to forge a collective good which emerges through processes of decision-making that are open, fair and take the interests of all members into account. The goal, then, is to design policies which ensure that the risks and hardships faced by members are distributed in a way that is as equitable as possible. At present, humans are the most powerful political animals on Earth; we should create and use institutions – from science to politics – to end the various systematic forms of oppression of animals, and to enable animals to have their say.

5. Conclusion

This paper has explored whether solidarity with wild animals is either present or possible. Through looking at three different forms of solidarity – social, political and civic – we have argued that solidarity with wild animals is not only present and possible, but an important concept to employ when thinking through our political relations with animals. Most crucially, employing the lens of solidarity to reflect on our relations with wild animals in particular, offers us a means to expand and transform our understanding of the political community in less anthropocentric and genuinely multi-species terms. By thinking about social solidarity, we can reflect on the communal identifications we already have with wild animals – as well as the ones we ought to have. By exploring political solidarity, we can think not only about the oppressive relations with wild animals that we ought to mobilize against and attempt to eradicate, but also the means by which animals' own perspectives can feed into that process. And thinking through civic solidarity with wild animals demands that we reflect on the institutional processes of our multi-species communities, and the ways in which the myriad interests of all members – human and non-human – can feed into constructions of the common good.

And focussing on solidarity in its various forms has certain other benefits. For example, it can also move discussions in animal ethics and politics beyond what we might call 'group-thinking'. At present, animals are often seen to belong to a specific group – that of domesticated, wild or liminal animals; accordingly, we debate their respective lots. Yet these groups are contingent and fluid, as can be demonstrated by the fact that some animals (like rabbits) 'occur' in all groups, or others (like feral goats) 'migrate' from one into another. And not only are these groups problematic in terms of categorizing animals, they are also problematic in terms of directing our political attention. For what seems important to our political relations with animals is less whether an animal is 'domesticated' or 'wild' but whether we identify communally with that animal, how we should identify with them given their contribution and participation in our societies, whether they are the victims of oppression, how we might respond to their needs at an institutional level, and so on.

Without doubt, important questions remain which have not been addressed in this paper. Is there a truly meaningful way in which solidarity could be extended to non-sentient beings (see Plumwood, 2001)? How can we draw the boundaries around multi-species communities to keep them both fluid and operational? And how can we form communities with members who do not necessarily form solidaristic relations with one another? While this paper leaves these and other important questions unanswered, we hope that it has done enough to show the importance of continued research into solidarity when thinking about membership and flourishing in our shared multi-species communities.

Notes

1. Martha Nussbaum (2006) takes seriously the idea of extending justice to wild animals, but does not go as far as suggesting that they might be fellow members of our communities.
2. It is certainly conceivable that certain wild animals, especially migratory ones, will have to be acknowledged by multiple political communities (just like many human beings).
3. Many wild animals are sociable *with each other*, of course.

Disclosure Statement

No potential conflict of interest was reported by the author(s).

ORCID

Mara-Daria Cojocaru http://orcid.org/0000-0001-6084-6380
Alasdair Cochrane http://orcid.org/0000-0002-3112-7210

References

Abrell, E. (2021). *Saving animals: Multispecies ecologies of rescue and care*. University of Minnesota Press.

Amir, F. (2021, November 5). *Pigeon politics*. https://evergreenreview.com/read/pigeon-politics/
BBC News. (2011). Greece's front-line riot dog. *BBC News*. Retrieved November 5, 2021, from https://www.bbc.co.uk/news/av/world-europe-13802940
BBC News. (2013). Nottinghamshire's police dogs to receive 'pensions'. *BBC News*. Retrieved July 19th, 2021, from https://www.bbc.co.uk/news/uk-england-nottinghamshire-24807719
Birke, L. (2009). Naming names — or, what's in it for the animals? *Humanimalia*, *1*(1), 1–9. https://doi.org/10.52537/humanimalia.10113
Bramble, B. (2021). Painlessly killing predators. *Journal of Applied Philosophy*, *38*(2), 217–225. https://doi.org/10.1111/japp.12461
Brookland, J. (2019, September 1). *Animal Welfare Commissioner, Labour's not so new idea*. Retrieved July 19, 2021, from https://animalrightsandwrongs.uk/tag/animal-ombudsman/
Browning, H., & Veit, W. (2021). *Positive wild animal welfare* (Preprint). Retrieved November 5th, 2021, from http://philsci-archive.pitt.edu/id/eprint/19608
Cochrane, A. (2018). *Sentientist politics: A theory of global inter-species justice*. Oxford University Press.
Cochrane, A., Garner, R., & O'Sullivan, S. (2018). Animal ethics and the political. *Critical Review of International Social and Political Philosophy*, *21*(2), 261–277. https://doi.org/10.1080/13698230.2016.1194583
Coulter, K. (2016). *Animals, work and the promise of interspecies solidarity*. Palgrave.
Donaldson, S., & Kymlicka, W. (2011). *Zoopolis: A political theory of animal rights*. Oxford University Press.
Eberle, U. (2021, June 23). *Can we use nudge theory to help endangered animals save themselves?* New Scientist. Retrieved July 19th, 2021, from https://www.newscientist.com/article/mg25033400-900-can-we-use-nudge-theory-to-help-endangered-animals-save-themselves/
Faria, C., & Paez, E. (2015). Animals in need. The problem of wild animal suffering and intervention in nature. *Relations*, *3*(1), 7–13.
Francione, G., & Charlton, A. (2015). *Animal rights: The abolitionist approach*. Exempla Press.
Garner, R. (2013). *A theory of justice for animals: Animal rights in a non-ideal world*. Oxford University Press.
Garner, R., & O'Sullivan, S. (Eds.) (2016). *The political turn in animal ethics*. Rowman and Littlefield.
Gould, C. (2007). Transnational solidarities. *Journal of Social Philosophy*, *38*(1), 148–164. https://doi.org/10.1111/j.1467-9833.2007.00371.x
Habermas, J. (1986). Gerechtigkeit und Solidarität. Eine Stellungnahme zur Diskussion über ›Stufe 6‹. In W. Edelstein & G. Nunner-Winkler (Eds.), *Zur Bestimmung der Moral. Philosophische und sozialwissenschaftliche Beiträge zur Moralforschung* (pp. 291–318). Suhrkamp.
Hediger, R., (Ed.). (2013). *Animals in war: Studies of Europe and North America*. Brill.
Hogan, B. (2020, March 3). *Is it possible to build wildlife-friendly windfarms?* BBC Future. https://www.bbc.com/future/article/20200302-how-do-wind-farms-affect-bats-birds-and-other-wildlife
Horta, O. (2010). Debunking the idyllic view of natural processes: Population dynamics and suffering in the wild. *Telos*, *17*(1), 73–88.
Hribal, J. (2011). *Fear of the animal planet: The hidden history of animal resistance*. AK Press.
Isfahani-Hammond, A. (2020, January 8). *How a Chilean dog ended up as a face of the New York City subway protests*. The Conversation. https://theconversation.com/how-a-chilean-dog-ended-up-as-a-face-of-the-new-york-city-subway-protests-129167
Jalagania, B. (2021). Wild animals and duties of assistance. *Journal of Agricultural & Environmental Ethics*, *34*(9). https://doi.org/10.1007/s10806-021-09854-3
Johannsen, K. (2020). *Wild animal ethics: The moral and political problem of wild animal suffering*. Routledge.
Lugones, M. (1987). Playfulness, "World"-Travelling, and loving perception. *Hypatia*, *2*(2), 3–19. https://doi.org/10.1111/j.1527-2001.1987.tb01062.x
Meijer, E. (2019). *When animals speak: Toward an interspecies democracy*. NYU Press.
Midgley, M. (1983). *Animals and why they matter: A journey around the species barrier*. Pelican.
Milligan, T. (2015). The political turn in animal rights. *Politics and Animals*, *1*(1), 6–15.
Moen, O. M. (2016). The ethics of wild animal suffering. *Etikk i Praksis - Nordic Journal of Applied Ethics*, *10*(1), 91–104. https://doi.org/10.5324/eip.v10i1.1972

Nussbaum, M. (2006). *Frontiers of justice: Disability, nationality, species membership*. The Belknap Press of Harvard University Press.

Pachirat, T. (2018). Sanctuaries. In L. Gruen (Ed.), *Critical terms for animal studies* (pp. 337–355). The University of Chicago Press.

Palmer, C. (2010). *Animal ethics in context*. Columbia University Press.

Palmer, C. (2012). The moral relevance of the distinction between domesticated and wild animals. In T. L. Beauchamp & R. G. Frey (Eds.), *The Oxford Handbook of animal ethics* (pp. 26). Oxford University Press. https://doi.org/10.1093/oxfordhb/9780195371963.013.0026

Palmer, C. (2016). Saving species but losing wildness: Should we genetically adapt wild animal species to help them respond to climate change? *Midwest Studies in Philosophy*, *XL*(1), 234–251. https://doi.org/10.1111/misp.12058

Pepper, A. (2021). Political agency in humans and other animals. *Contemporary Political Theory*, *20*(2), 296–317. https://doi.org/10.1057/s41296-020-00430-4

Plumwood, V. (2001). *Environmental culture: The ecological crisis of reason*. Routledge.

Prainsack, B., & Buyx, A. (2012). Solidarity in contemporary bioethics – toward a new approach. *Bioethics*, *27*(7), 343–350. https://doi.org/10.1111/j.1467-8519.2012.01987.x

Regan, T. (2004). *The case for animal rights* (2nd ed.). University of California Press.

Rippe, K. (1998). Diminishing solidarity. *Ethical Theory and Moral Practice*, *1*(3), 355–374. https://doi.org/10.1023/A:1009965816147

Rock, M., & Degeling, C. (2015). Public health ethics and more-than-human solidarity. *Social Science and Medicine*, *129*(March), 61–67. https://doi.org/10.1016/j.socscimed.2014.05.050

RSPB. (2021, January 29). *Two thirds of UK found solace in watching birds and hearing their song over lockdown, survey reveals ahead of Big Garden Birdwatch*. RSPB News. https://www.rspb.org.uk/our-work/rspb-news-original/news/stories/birdwatch/

Scholz, S. (2008). *Political solidarity*. Pennsylvania State University Press.

Scholz, S. (2013). Political solidarity and the more-than-human world. *Ethics & the Environment*, *18*(2), 81–99. https://doi.org/10.2979/ethicsenviro.18.2.81

Singer, P. (1972). Famine, affluence, and morality. *Philosophy and Public Affairs*, *1*(3), 229–243.

Singer, P. (1995). *Animal liberation* (2nd ed.). Pimlico.

Smith, K. (2012). *Governing animals: Animal welfare and the liberal state*. Oxford University Press.

Vink, J. (2020). *The open society and its animals*. Palgrave.

von Essen, E., & Allen, M. P. (2020). Interspecies political agency in the total liberation movement. *Between the Species*, *23*(1), Article 7. https://digitalcommons.calpoly.edu/bts/vol23/iss1/7/

Wissenburg, M., & Schlosberg, D. (2014). *Political animals and animal politics*. Palgrave.

WWF. (n.d.). *Gifts in wills FAQs*. https://www.wwf.org.uk/faqs/gifts-in-wills

Young, I. M. (1990). *Justice and the politics of difference*. Princeton University Press.

Reducing Wild Animal Suffering Effectively: Why Impracticability and Normative Objections Fail Against the Most Promising Ways of Helping Wild Animals

Oscar Horta and Dayron Teran

ABSTRACT
This paper presents some of the most promising ways wild animals are currently being helped, as well as other ways of helping that may be implemented easily in the near future. They include measures to save animals affected by harmful weather events, wild animal vaccination programs, and projects aimed at reducing suffering among synanthropic animals. The paper then presents other ways of helping wild animals that, while noncontroversial, may reduce aggregate suffering at the ecosystem level. The paper argues that impracticability and normative objections against reducing wild animal suffering fail against these ways of helping wild animals.

There are many different ways in which wild animals are currently helped by human beings. Some ways of helping aid just a few animals and are relatively common. They include, among many others, cases when people acting individually rescue animals drowning in ponds or mud ponds, trapped in frozen lakes, stranded in beaches, and affected by floods, fires and different types of natural disasters. Some people build structures that shelter animals from the effects of severe weather and other threats. At a larger scale, some private or public initiatives help a greater number of animals. These include rescue centers, wild animal hospitals and orphanages, where sick, injured or orphaned animals who would otherwise probably die are taken care of. In some cases, feeding projects conducted during especially harsh winters, or during droughts, save the lives of many starving animals. Finally, other projects help animals much more extensively. The clearest example may be wild animal vaccination programs (Anderson & Anderson, 2006; Animal Ethics, 2020a, ch. 8–10; Bovenkerk et al., 2003; Delahay et al., 2009; Kirkwood & Sainsbury, 1996; Marion et al., 2008; Thompson et al., 2015).

Wild animals could be helped in other, more significant ways if there were a will to do so. This would require investing the resources needed to gain more knowledge about the best ways to help. There are clear examples that show how this could be done. For the sake of both biological conservation and human interests, a very significant amount of resources has been invested in researching the ways ecosystems work. The interventions implemented to pursue these two aims have been informed by the knowledge acquired by such research. Similarly, research could be done for the sake of reducing wild animal suffering. The progress made thus far in achieving this aim has been very little, especially in comparison to anthropocentric and conservationist goals. But this has not been due to there being something especially inscrutable or intractable about reducing such suffering, but because no significant resources have been invested in the field.

Part of the explanation for why wild animal suffering has been neglected is simply that animals are still largely disregarded in our societies. Another reason is that it is often taken for granted that animals have happy lives in the wild and that there is nothing they need help with. This idyllic view is, however, challenged by the above-mentioned examples of ways in which animals are rescued from situations of dire need. Such cases take place in circumstances where animals are gravely threatened and are often undergoing severe suffering. In fact, the available evidence indicates that such cases are not uncommon, given the many factors negatively affecting wild animals' wellbeing (Animal Ethics, 2020a; Duclos, 2022; Faria, 2023; Horta, 2017; Horta, 2017/2022; Johannsen, 2020; Ng, 1995; Tomasik, 2010/2015). Such factors can pose formidable challenges that cause wild animals to die in painful ways, often shortly after coming into existence and without having had the chance to enjoy significant pleasurable experiences. This, together with the large numbers of wild animals that there are (Ray, 2017; Tomasik, 2009/2019), means that efforts aiming at helping them are much more important than they might seem at first glance.

In light of its scale, there has been a growing amount of literature in recent years claiming that wild animal suffering should receive much more attention than it currently does. In addition to expressing support for existing ways of helping wild animals, the most common claim made by those who have worked on this problem is that more research should be dedicated to it (see for instance Animal Ethics, 2020a; Faria & Horta, 2019; Ng, 1995; Soryl et al., 2021). But these claims have met with objections. Some of them are what we can call impracticability or intractability objections. They appeal mostly to epistemic limitations in order to argue that wild animal suffering is an intractable cause area. Other objections, instead, are normative ones, as they are based on the idea that helping wild animals is not desirable. This paper will use the information available about wild animal suffering to argue that such objections fail, especially considering the ways of helping animals that are currently being implemented, and the ways that may be implemented in the near future.

This paper will first present several ways of helping wild animals that are especially promising at this point because of their potential to foster future work on the topic. In doing this, it will consider the results of a recent survey on this topic among natural science scholars and students done by the animal organization Animal Ethics (2020b). The paper will then explain the kind of novel, cross-disciplinary scientific work that would be necessary in order to increase our understanding of the factors negatively affecting wild animals, and of the ways to reduce their suffering. It will then consider the objection that there is nothing that can be done to effectively reduce wild animal suffering. The paper will argue that the available ways of helping wild animals suggest this objection does not present a serious challenge. Next, it will argue that the ways of helping wild animals highlighted in the survey by Animal Ethics can avoid some of the main normative objections against preventing wild animal suffering. Finally, the paper concludes by indicating some current perspectives on the development of this field of work.

Promising Ways of Helping Wild Animals

A survey published in 2020 by the animal advocacy organization Animal Ethics (2020b) identified three areas of research considered to be especially promising by biology, environmental studies and veterinary science scholars and students: helping animals affected by weather events, wild animal vaccination and helping wild animals in urban environments. Over 80% of respondents among both scholars and students agreed that research on those topics would provide useful knowledge (research on helping animals in urban areas was considered especially promising, with more than 90% of respondents expressing support for it).[1] In addition, the majority of scholars (in some cases, again over 80%) also agreed that such research would receive support from fellow scholars and from their university, and the majority of students agreed that they would be supported by their professors and by their fellow students. It thus looks like these three topics may be promising topics to start working on. We will see now what work in each of them consists of.

Rescuing Animals Affected by Extreme Weather Events

Weather events regularly affect animals negatively, sometimes in very serious ways. For instance, many animals drown in floods or due to lake thaws, or are removed from the areas where they live by floods or by heavy winds. Many animals that survive these events get orphaned and cannot fend by themselves. Extreme weather can also cause other events that can severely harm very large numbers of animals, such as forest fires.

Concerned individuals sometimes aid animals in these situations. But, at a larger scale, protocols aiming at preventing the harms that animals can suffer during and after these events, or at introducing and improving the efficiency of rescue programs carried out during these event, can be designed and implemented (Gutiérrez & de Miguel, 2021; McCumber & King, 2020; Sássi et al., 2021; Sawyer & Huertas, 2018).

Vaccination

We already mentioned the importance of wild animal vaccination programs. These have been carried out for several decades already in different parts of the world, targeting diseases such as anthrax, brucellosis, tuberculosis and, especially, rabies, among others (Gormley et al., 2017; Rupprecht et al., 2004; Turnbull et al., 2004). A common method for delivering the vaccine is the use of oral baits that the targeted animals are attracted to. To cover wide areas, these baits can be air-dropped. In other cases, other delivery methods are used: for instance, bats have been vaccinated using sprays that reach them in large numbers (Fontaine, 2017).

While vaccination programs are implemented with the aim of preventing diseases from being passed to either animals used by humans or to humans themselves, these programs have nonetheless been very beneficial for the very large numbers of nonhuman animals who have been immunized. Rabies, for instance, has been eradicated from large parts of different countries. The success of wild animal vaccination programs, coupled with the importance that they may have to human health, means there is currently a lot of interest in continuing to invest in them. Furthermore, interest in these programs is likely to increase further due to the COVID-19 pandemic. The prevention of future outbreaks caused not just by new coronaviruses different from SARS-CoV-2, but also by other pathogens, is likely to receive more attention (Edwards et al., 2021; Monath, 2013). In this way, human interest in avoiding future zoonotic diseases will as a side effect continue to reduce suffering and deaths among wild animals.

Reducing Suffering Among Synanthropic Animals

When people first think of wild animals, they tend to think of animals living in the wilderness. However, many wild animals live close to humans, in environments that have been transformed very significantly by humans. These include urban, suburban, industrial and agricultural areas. Animals living there are often called synanthropic, which means that they live close to humans. Synanthropic animals include, among others, many types of birds, small mammals and reptiles, amphibians, and arthropods. In some areas, they can include larger animals like foxes, storks, iguanas, etc. (Adams, 2016; Burger & Lunney, 2004; Hadidian & Smith, 2001; Michelfelder, 2018).

These animals face different opportunities and challenges. They not only face a number of anthropogenic threats, but also suffer due to natural factors. Due to the latter, a good starting point for measures that aim to reduce wild animal suffering could be to implement programs that aim to improve the lives of synanthropic animals. Programs would be easier to implement, monitor, and assess in the areas where synanthropic animals live than in more remote places. Such programs could also include assistance programs for animals affected by extreme weather, and vaccination programs. Protocols for extreme weather events (and other natural disasters) affecting urban and rural areas are increasingly taking domesticated animals into consideration. They could take wild animals into consideration as well. As for wild animal vaccination, it seems that vaccination plans for synanthropic animals are easy to defend for anthropocentric reasons alone, and are a good way to start increasing concern for these animals.

Other Promising Areas of Work

In addition to the ways of helping animals identified in the survey mentioned above, there are others that seem to be promising for at least three reasons: they appear to be relatively noncontroversial, their development can take advantage of research that has already been done, and they may have great potential in the long run for helping a large number of animals. We will consider two of them here.

Contraception

Wild animal immunocontraception methods have been developed in recent decades, targeting especially mammals and birds (Ansari et al., 2017; Asa et al., 2012; Cooman et al., 2011; Patton et al., 2007). While in some cases the administration of contraceptives requires interacting with the animals (as in the case of injectable GonaCon immunocontraceptive vaccines for ungulates), in others this is not necessary, since the contraceptives can be delivered using oral baits (Cohn & Kirkpatrick, 2015; Fagerstone et al., 2010).

Contraception can save many animals from suffering and dying from both anthropogenic and natural causes. It sometimes happens that certain animal populations grow larger than humans want them to be, and as a result, members of those populations are hunted down. We can of course question the reasons for why this is done. However, we may not be able to convince the relevant decision-makers that their reasons are poor. In those cases, we can defend contraception as an alternative approach that would save the lives of animals who would otherwise be killed. But this is not the only way contraception can help animals. It can also help to prevent animal populations from growing to the point where their members die because they do not have enough food or other resources in their environments. Moreover, contraception can be what eventually makes it possible for other measures helping animals to succeed. In fact, this may be where its most significant potential lies in the long-term. To see why, consider an example. Suppose that there is a certain population of animals whose members are suffering due to a certain factor limiting its growth, such as, say, a disease. Suppose, further, that we have the opportunity to eradicate this disease through some treatment. However, if we do so, there is a risk that the population will grow substantially and that, in the end, we will merely be postponing its members' suffering. A solution may consist in combining the treatment for the disease with a contraception program. This may allow us to improve the present situation of this population, while preventing population growth and the problems it would cause. Contraception may be even more important in cases where animals are provided supplemental feeding, since feeding programs have an even greater risk of increasing the populations they target.

Measures Reducing Aggregate Suffering at the Ecosystem Level

There are other types of measures that can have an impact not just in certain individuals or populations, but at the ecosystem level. Consider, for instance, decisions concerning whether a previously existing ecosystem should be restored or not, or about what kind of trees or bushes should be planted if a reforestation plan is going to be implemented. In such cases, we may be able to make some rough predictions about the different ecosystems that different decisions would produce in the targeted area. Accordingly, we may be able to guess what kind of animals will be there in each case. Such animals might be among those who have higher survival rates and longer average lifespans, and who reproduce by having small numbers of offspring. Or they may, instead, be among those who reproduce in very large numbers and tend to die in their very early youth. The latter, who unfortunately are the majority in nature, typically have much harder lives. Their lives may be so hard that they often contain more suffering than pleasure (Faria & Paez, 2015; Horta, 2017; Ng, 1995).

In this way, it's possible to roughly compare the expected wellbeing of the animals who would exist in different possible ecosystems. Accordingly, we may reach the conclusion that in one of them animals will generally be in a better situation. If that is so then we would have reasons to favor the presence of that ecosystem, instead of alternative ones where animals do worse. In particular, we would have reason to favor the ecosystems containing animals that tend to have longer lifespans, rather than those containing animals who typically die shortly after coming into existence. The expected wellbeing of animals could be considered in our decisions concerning whether a reforestation program should take place in one way or in another way. It could also be used to decide if a certain area should be reforested at all. If the ecosystem existing at present in that area has a lower level of wellbeing for animals than the one that would result from reforestation, then we should favor reforestation, although if it were the other way around, we would have reason to oppose reforestation. (These comparisons can be done regardless of the particular philosophical conception of wellbeing we hold).

A Cross-Disciplinary Applied Approach

The kind of approach needed in order to successfully research ways of helping animals like the ones listed above is one that combines the methods and viewpoints of different disciplines. To be more specific, successfully researching how to help wild animals would require two things: (i) expanding animal welfare assessments to wild animals living outside direct human control (JWD Wildlife Welfare Supplement Editorial Board, 2016; Kirkwood & Sainsbury, 1996); and (ii) looking at the knowledge that can be gathered in different fields in the science of ecology, and related disciplines, through the eyes of the science of animal welfare (Faria & Horta, 2019; Soryl et al., 2021). While novel, this cross-disciplinary approach – sometimes referred to as 'welfare biology' (Ng, 1995) – would nevertheless be grounded on the knowledge previously gathered in those fields.

Methods of welfare assessment have developed significantly during the last decades, although for the most part they have been applied to domesticated animals of the most populated species (mainly those that are kept and killed for the production of food products, such as pigs, cows and chickens, among others). Similar assessments of other animals have been done only to a very limited extent. But progress in this field can be made with respect to animals in the wild as well. This would provide us with very useful insights into how wild animals are feeling. In addition to developing these assessments it would be necessary to gather more knowledge about the different external factors affecting wild animals' wellbeing. This would be a job for the science of ecology. Due to a disregard for animals as individuals with a wellbeing of their own, previous work in ecology has not focused on the experiences animals have, but on other matters, such as the circumstances in which animals find themselves, how they try to deal with them, and the manner in which some animals are eventually able to transmit their genetic information to further generations. But such work can nevertheless be enormously useful for making informed evaluations of how animals in the wild are harmed in different ways. This is because such work has included, to a large extent, an examination of the factors affecting animal populations, such as weather variations, food and water availability, antagonism with other animals, disease, etc. The impact of these factors has been measured extensively in many different populations all around the world. Such measurements have been made in terms of things such as mortality, reproductive success, health indicators, and other dimensions different from wellbeing. However, they provide significant information about the wellbeing of animals. The reason is that when animal populations are affected by a certain factor in ways that increase mortality and reduce the average health of animals, it is reasonable to infer that the wellbeing of those animals is also being negatively affected (Faria & Horta, 2019; Hecht, 2021). For instance, we know a great deal already about the impact of variations in temperature on animal populations (Bowler et al., 2017; Dillon et al., 2016). But if a cold wave kills half of a population, we can conclude that the animals that were killed were harmed by their death (Bradley, 2016; Kaldewaij, 2006). In addition, it makes sense to assume that they suffered in the process of dying. Moreover, it seems implausible that the only animals who suffered from the cold wave were the ones who died. Rather, the cold affected all the animals in the population, so it is likely that the rest of those animals also had their wellbeing worsened, to different degrees, by the cold wave. Similar inferences can be made when we consider other factors.

In addition, the know-how already possessed in the fields of applied ecology and wildlife management can be used to implement measures benefiting animals. Work designing actual programs that put into practice the ways of helping animals listed above can be useful not just in themselves, but because they can light the path to further cross-disciplinary work that would develop this novel field.

Responding to Intractability Objections

Having considered the positive case for expanding efforts to help wild animals, we can now consider objections to such efforts. Perhaps the most common objection is that, due to the complexity of ecosystems, we do not know all the possible effects that our actions could have on them. There could be a serious risk of causing unforeseen catastrophic outcomes. Therefore, the objection goes, for the sake of animals themselves, it would be safer to leave things as they are, since in trying to help animals we may make things worse for them (Delon & Purves, 2018).

There are several responses that can be given to this objection. The following ones may be the most relevant:

Stressing the Importance of Research

The first point to make here is something we have stressed throughout this paper, especially in the previous section. We have stressed the importance of developing the study of the wellbeing of wild animals, the factors affecting them, and the ways to help them. Gathering this knowledge may have a much higher impact in the long run than rushing to act immediately. To be sure, in order to optimize the results of our actions we should try to minimize risks. But if we give up on helping wild animals, we will not just minimize risk, but abandon altogether the goal of having a positive impact. Instead, fostering further research can help to attain this goal while significantly reducing risks.

The Example of Standard Practice in Ecology

In addition, the objection is at odds with the way we proceed when human interests or biological conservation are at stake. When interventions in ecosystems are implemented for conservationist or anthropocentric reasons, scientists do not expect to have complete knowledge of everything that will happen. Yet they make some reasonable predictions on the basis of a great deal of accumulated knowledge. Current practice in ecology, while certainly supportive of the idea that ecosystems are very complex and difficult to predict, also assumes that the success of programs in applied ecology is possible so long as they are adequately planned and monitored. On the basis of this, scientists accept some reasonable risks for the sake of such goals, in part because inaction is costly. We should do the same for the sake of animals' interests. Of course, conservation biologists typically study, in great detail, the problem of unforeseen consequences, and on the basis of this they decide to limit their activities accordingly, or to go ahead with them. Welfare biologists should proceed in the same way.

It could be argued that aiming to reduce the suffering of wild animals is problematic, while aiming to achieve other goals within ecosystems is not. But it is not clear that this particular aim is uniquely problematic. To be sure, it could be argued that it is very difficult to know exactly what kind of mental states nonhuman animals have. However, this objection does not seem to be correct, for two reasons. First, as we have seen above, even if we can never access animals' mental states, there are indicators we can use – we have seen already how death can be used as a proxy for suffering. Second, this objection points to something important, but different from the matter at stake here. The problem we are examining is whether efforts that aim to reduce wild animal suffering should be expected to fail because there will always be unexpected consequences. The basis for claiming this is not the difficulty of adequately assessing the well-being of nonhuman animals. Rather, it is the complexity of ecosystems, which is a factor that is also relevant to interventions carried out for the sake of human interests or biological conservation.

A Comparison between Our Assessment of Risks and of the Present Situation

Furthermore, it is worth pointing out that, if we accept that a certain action may have unexpected effects, we cannot necessarily assume that they would be bad, since by definition we do not know what they would be like. They may be negative, neutral or positive. Moreover, we should also bear in mind that by refraining from helping animals we are not just avoiding the risk of a bad outcome; rather, we are letting wild animal suffering take place. Due to this, intractability objections may not only be overestimating how negative the potential consequences of helping wild animals might be; but also undervaluing how negative the present situation is for wild animals.

In relation to this, it can be argued that in objecting to helping wild animals, we may be giving priority to the present situation just because it happens to be the present situation. That is, as a result of a status quo bias, and not because the present situation is necessarily better than the outcome that will occur if we take action. One way to discern if this is so is by applying the reversal test against this bias to different actions implemented to help wild animals (Bostrom & Ord, 2006; Faria, 2023). For instance, suppose someone opposes a wild animal vaccination program that will prevent large numbers of animals from suffering, and eventually dying, from a certain disease. The reason for such opposition would be that improving the immunity of the targeted animals is either bad in itself or would lead to bad consequences. To apply the reversal test, we would have to consider decreasing the immunity of the targeted animals. Would decreasing their immunity be good or bad? It seems the response would be negative here. Even if we do not think that the reverse intervention, that is, decreasing the animals' immunity, is bad in itself, we can easily reach the conclusion that it would lead to bad consequences, as it would mean that more of these animals will suffer and die. But then, that would mean that the current state is at an optimal level for animals in the wild, since no matter what change we make to the present situation, the result would be worse. However, it would be very coincidental if, of all the possible situations wild animals could be in, they just happened to be in the best one at present. The same conclusion can be reached about other ways of helping animals. In fact, the general idea that animals in the wild just happen to be in the best possible situation seems implausible. Hence, the reversal test suggests that we are indeed biased against trying to improve the present situation of animals (Faria, 2023).

Current Practice and Scientists' Judgment

Finally, skepticism about the possibility of helping wild animals contradicts the fact that people all around the world are already helping wild animals in a number of situations, sometimes at an institutional level and targeting very large numbers of animals. Moreover, as the previously discussed Animal Ethics survey shows, this skepticism does not match the opinions of scientists. Both scientists and science students had a positive opinion about the usefulness of further researching the mentioned ways of helping wild animals. In fact, the results were quite overwhelming, as a huge majority held this opinion (over 80% to 90%, depending on the case). Skeptics may argue that the respondents' judgment should be doubted, as the fact that many people agree with something does not mean that they are right. However, this objection would apply equally to the opinion that trying to reduce wild animal suffering is too risky. In light of this, we may be led to think that we are uncertain here and have no reason to agree with either those who support helping wild animals or those who reject doing so. However, these opinions are not epistemically equal. Scientists and, to a lesser extent, students in fields like biology, veterinary science and environmental sciences are much more familiar with the different ways that humans intervene in nature, and with assessing the risks involved in doing so, than most other people are. Due to this, their judgment seems to be more reliable. In fact, because the prevalent value paradigms in natural sciences exclude the moral consideration of animals, and are based instead on anthropocentric and conservationist ideas, if anything we should expect the judgment of scientists and science students to be somewhat biased against the idea of helping animals. The fact that, despite this, scientists and science students have expressed a positive view toward researching ways of helping wild animals, suggests that the scientific community may have important reasons to doubt impracticability objections.

If these responses to intractability objections are right, then it is possible for efforts to reduce wild animal suffering to succeed. It may be argued, however, that such efforts will not be able to eventually end it. As long as there are ecosystems with sentient organisms in them like the ones existing today, large amounts of suffering will continue to be present in them. Skeptics about helping wild animals may thus argue that efforts toward this cause are futile (Delon & Purves, 2018), and that addressing problems other than wild animal suffering is therefore more important because those problems can be solved completely while wild animal suffering cannot. But this conclusion does not follow. The relevant question is not whether it is possible to eliminate all suffering, or even all extreme suffering. What is relevant is the extent to which it is possible to reduce such suffering and therefore to improve the wellbeing of animals. We should avoid the futility thinking bias according to which if we cannot solve a certain problem completely then working on it is not worth it, even when we can indeed solve it at least partly (Unger, 1996 see also Faria, 2023). Because of the vast scale of wild animal suffering, solving a tiny part of it can improve the situation of a very large number of animals. This can mean achieving a total improvement to wellbeing that is much higher than the one we could achieve by completely solving problems that affect smaller numbers of animals or human beings.

Responding to Normative Objections

Other objections are normative. They challenge not whether we can succeed in helping wild animals, but whether doing so is desirable. For instance, according to holist objections, we should not care about what happens to individual animals, but rather to wholes like ecosystems or species (Callicott, 1989; Sagoff, 1984). According to wilderness-focused, naturocentric views in environmental ethics, we should just care for the preservation of wild areas (Katz, 1992; Elliot, 1997; These views are examined in Duclos, 2022). According to speciesist objections, we should not help animals because we should care about human interests alone (Carruthers, 1992; Narveson, 1977). Finally, there are also views according to which it is wrong to reduce any harms animals suffer that are not anthropogenic (Everett, 2001; see also Morris & Thornhill, 2006).

All these views may be rejected if we accept that we should be concerned with the interests of all sentient beings, regardless of their species and of the circumstances affecting them (Faria, 2023; Sapontzis, 1987; Torres, 2015). However, in this paper we are not going to argue against these views. Rather, we will argue that even those who accept them have no reason to oppose at least some ways of helping wild animals.

Holism is at least compatible with helping synanthropic animals, since they do not live in ecosystems typically valued by holistic views. Helping synanthropic animals is also compatible with wilderness-focused views, since urban, industrial and agricultural environments are no longer natural. In fact, wilderness-focused views are compatible with helping animals in other ecosystems created by humans, such as pasture lands, forests planted for wood production, and parks managed for recreation outside urban areas (Horta, 2017, 2018).

For their part, anthropocentric speciesist views can support ways of helping animals that are beneficial for humans (and may not even reject other ways of helping if enough people support them). Ways of helping that benefit humans include, for instance, wild animal vaccination and contraception programs. The development of these programs would be useful, too, for developing interventions motivated by non-anthropocentric concerns.

Finally, the objection that we should only help animals affected by anthropogenic harms has significantly less importance than it may seem to at first. To start with, it should not oppose efforts directed at helping synanthropic animals, since they live in areas continually transformed by humans and are regularly affected, positively or negatively, by human action. Furthermore, the objection does not apply within the context of ecosystems created by human beings, such as the ecosystems we saw when we discussed measures that reduce aggregate suffering at the ecosystem level. Examples of these new ecosystems include those resulting from reforestation programs, or from other initiatives like desert greening. Such ecosystems end up being populated by many animals who would otherwise not have existed (especially when the new ecosystem's primary production is higher than that of the previously existing ecosystem). These animals face many challenges, a result of which is that many of them suffer significantly and die prematurely. These harms would not have taken place without human action. Accordingly, measures that aim to reduce the suffering of animals in anthropogenic ecosystems are unaffected by the objection that nonanthropogenic harms should not be prevented. In addition, since anthropogenic climate change is currently the most likely cause of many extreme weather events, views concerned with anthropogenic harms would have reason to support helping animals affected by such events (Palmer, 2021; Sebo, 2022). In fact, since humans' impact on climate has, in one way or another, affected all ecosystems around the world, the distinction between natural and anthropogenic harms is becoming blurry.

Conclusion

This paper has argued that the claim that wild animal suffering is a hopeless cause is rebutted by the available evidence. Wild animals are currently being helped in many circumstances. Further work in this cause area could very significantly expand these efforts. Cross-disciplinary research on the factors negatively affecting the wellbeing of animals can play a crucial role in achieving this expansion. In this paper we have pointed to several ways of helping animals that have a high potential for increasing the attention and resources directed to this kind of research. Promoting more work on these ways of helping looks like a promising strategy for developing the cause area of reducing wild animal suffering.

Despite being impactful, these ways of helping animals are realistic, as well as rather noncontroversial. Due to this, they avoid intractability objections, and they also avoid normative ones. People defending these objections may disagree with some ways of helping animals that may be implemented in the future, such as ways involving significant ecological redesign and the use of gene drive technology. But they should have no issues with ways of helping that wild animal advocates have reason to foster right now. This is good news, as it means that in effectively reducing wild animal suffering, a lot of progress can currently be made that need not be met with opposition.

Note

1. The main results of the survey were as follows (see Animal Ethics, 2020b for more details).

   92% of scholars agreed that work on urban animals' wellbeing would produce important and useful knowledge. 89% of them agreed in the case of wild animal vaccination; and 85% agreed in the case of helping wild animals harmed by weather events. Among students, 93% agreed in the case of work on urban animals; 89% agreed in the case of vaccination; and 81% agreed in the case of animals in weather events.

   85% of scholars agreed that work on urban animals would be supported by fellow scientists. 80% of them agreed in the case of vaccination; and 74% agreed in the case of animals in weather events. Among students, 81% agreed that work on urban animals would be supported by their professors, 72% agreed in the case of vaccination; and 71% agreed in the case of animals in weather events. In addition, 86% agreed that work on urban animals would be supported by fellow students, 82% agreed in the case of vaccination; and 77%

agreed in the case of animals in weather events.

Finally, 62% of scholars agreed that work on urban animals would be supported by their university departments. 56% of them agreed in the case of vaccination; and 73% agreed in the case of animals in weather events.

Acknowledgment

We thank Kyle Johannsen for helping to improve this paper with his comments and suggestions.

Disclosure Statement

No potential conflict of interest was reported by the author(s).

ORCID

Oscar Horta http://orcid.org/0000-0002-1001-9632
Dayron Teran http://orcid.org/0000-0002-7928-6472

References

Adams, C. E. (2016). *Urban wildlife management*. CRC Press.

Anderson, A., & Anderson, L. (2006). *Rescued: Saving animals from disaster*. New World Library.

Animal Ethics. (2020a). *Introduction to wild animal suffering: A guide to the issues*. Animal Ethics. Retrieved June 2021, from https://www.animal-ethics.org/introduction-wild-animal-suffering

Animal Ethics. (2020b). *Surveying attitudes toward helping wild animals among scientists and students*. Animal Ethics. Retrieved June 2021, from https://www.animal-ethics.org/survey-helping-wild-animals-scientists-students

Ansari, A. S., Ayesha, B., & Lohiya, N. K. (2017). Fertility control modalities in animals: An overview. *BAOJ Veterinary Science*, *1*(1), 004.

Asa, C., Boutelle, S., & Bauman, K. (2012). AZA wildife contraception center programme for wild felids and canids. *Reproduction in Domestic Animals*, *47*, 377–380. https://doi.org/10.1111/rda.12004

Bostrom, N., & Ord, T. (2006). The reversal test: Eliminating status quo bias in applied ethics. *Ethics*, *116*(4), 656–679. https://doi.org/10.1086/505233

Bovenkerk, B., Stafleu, F., Tramper, R., Vorstenbosch, J., & Brom, F. W. A. (2003). To act or not to act? Sheltering animals from the wild: A pluralistic account of a conflict between animal and environmental ethics. *Ethics, Place & Environment*, *6*(1), 13–26. https://doi.org/10.1080/13668790303539

Bowler, D. E., Haase, P., Hof, C., Kröncke, I., Baert, L., Dekoninck, W., Domisch, S., Hendrickx, F., Hickler, T., Neumann, H., O'Hara, R. B., Sell, A. F., Sonnewald, M., Stoll, S., Türkay, M., Klink van, R., Schweiger, O., Vermeulen, R., & Böhning-Gaese, K. (2017). Cross-taxa generalities in the relationship between population abundance and ambient temperatures. *Proceedings of the Royal Society B: Biological Sciences*, 284, 20170870.

Bradley, B. (2016). Is death bad for a cow? In T. Višak & R. Garner (Eds.), *The ethics of killing animals* (pp. 51–64). Oxford University Press.

Burger, S. K., & Lunney, D. (Eds.). (2004). *Urban wildlife: More than meets the eye*. Royal Zoological Society of New South Wales.

Callicott, J. B. (1989). *In defense of the land ethic: Essays in environmental philosophy*. State University of New York Press.

Carruthers, P. (1992). *The animal issue: Moral theory in practice*. Cambridge University Press.

Cohn, P., & Kirkpatrick, J. F. (2015). History of the science of wildlife fertility control: Reflections of a 25-year international conference series. *Applied Ecology and Environmental Sciences*, *3*(1), 22–29. https://doi.org/10.12691/aees-3-1-5

Cooman, L. D., Garmyn, A., Waeyenberghe, L. V., & Martel, A. (2011). Contraception in birds. *Vlaams Diergeneeskundig Tijdschrift, 80*(3), 201–213.

Delahay, R. J., Smith, G. C., & Hutchings, M. R. (2009). *Management of disease in wild mammals*. Springer.

Delon, N., & Purves, D. (2018). Wild animal suffering is intractable. *Journal of Agricultural & Environmental Ethics, 31*(2), 239–260. https://doi.org/10.1007/s10806-018-9722-y

Dillon, M. E., Woods, H. A., Wang, G., Fey, S. B., Vasseur, D. A., Telemeco, R. S., Marshall, K., & Pincebourde, S. (2016). Life in the frequency domain: The biological impacts of changes in climate variability at multiple time scales. *Integrative and Comparative Biology, 56*(1), 14–30. https://doi.org/10.1093/icb/icw024

Duclos, J. (2022). *Wilderness, morality, and value*. Lexington Books.

Edwards, S. J., Chatterjee, H. J., & Santini, J. M. (2021). Anthroponosis and risk management: A time for ethical vaccination of wildlife? *The Lancet Microbe, 2*(6), e230–231. https://doi.org/10.1016/S2666-5247(21)00081-1

Elliot, R. (1997). *Faking nature: The ethics of environmental restoration*. Routledge.

Everett, J. (2001). Environmental ethics, animal welfarism, and the problem of predation: A Bambi lover's respect for nature. *Ethics & the Environment, 6*(1), 42–67. https://doi.org/10.1353/een.2001.0005

Fagerstone, K. A., Miller, L. A., Killian, G., & Yoder, C. A. (2010). Review of issues concerning the use of reproductive inhibitors, with particular emphasis on resolving human–wildlife conflicts in North America. *Integrative Zoology, 5*(1), 15–30. https://doi.org/10.1111/j.1749-4877.2010.00185.x

Faria, C. (2023). *Animal ethics in the wild: Wild animal suffering and intervention in nature*. Cambridge University Press.

Faria, C., & Horta, O. (2019). Welfare biology. In B. Fischer (Ed.), *Routledge handbook of animal ethics* (pp. 455–466). Routledge.

Faria, C., & Paez, E. (2015). Animals in need: The problem of wild animal suffering and intervention in nature. *Relations: Beyond Anthropocentrism, 3*(1), 7–13.

Fontaine, N. (2017). *Probiotic application of symbiotic bacteria isolated from western bat species onto captive and field bats to prevent white-nose syndrome* [Master thesis]. Thompson Rivers University.

Gormley, E., Ní Bhuachalla, D., O'Keeffe, J., Murphy, D., Aldwell, F. E., Fitzsimons, T., Stanley, P., Tratalos, J. A., McGrath, G., Fogarty, N., Kenny, K., More, S. J., Messam, L. L. M., & Corner, L. A. L. (2017). Oral vaccination of free-living badgers (Meles meles) with Bacille Calmette Guérin (BCG) vaccine confers protection against tuberculosis. *PLoS One, 12*(1), e0168851. https://doi.org/10.1371/journal.pone.0168851

Gutiérrez, J., & de Miguel, J. (2021). Fires in nature: Review of the challenges for wild animals. *European Journal of Ecology, 7*(1), 95–117. https://doi.org/10.17161/eurojecol.v7i1.14643

Hadidian, J., & Smith, S. (2001). Urban wildlife. In D. J. Salem & A. N. Rowan (Eds.), *The state of the animals* (pp. 165–182). Humane Society Press.

Hecht, L. (2021). The importance of considering age when quantifying wild animals' welfare. *Biological Reviews, 96*(6), 2602–2616. https://doi.org/10.1111/brv.12769

Horta, O. (2017). Animal suffering in nature: The case for intervention. *Environmental Ethics, 39*(3), 261–279. https://doi.org/10.5840/enviroethics201739320

Horta, O. (2017/2022). *Making a stand for animals*. Routledge.

Horta, O. (2018). Concern for wild animal suffering and environmental ethics: What are the limits of the disagreement? *Les ateliers de l'éthique, 13*(1), 85–100. https://doi.org/10.7202/1055119ar

Johannsen, K. (2020). *Wild animal ethics: The moral and political problem of wild animal suffering*. Routledge.

JWD Wildlife Welfare Supplement Editorial Board. (2016). Advances in animal welfare for free-living animals. *Journal of Wildlife Diseases, 52*(2s), S4–SS13. https://doi.org/10.7589/52.2S.S4

Kaldewaij, F. (2006). Animals and the harm of death. In M. Kaiser & M. Lien (Eds.), *Ethics and the politics of food* (pp. 528–532). Wageningen Academic Publishers.

Katz, E. (1992). The call of the wild: The struggle against domination and the technological fix of nature. *Environmental Ethics, 14*(3), 265–273. https://doi.org/10.5840/enviroethics199214321

Kirkwood, J. K., & Sainsbury, A. W. (1996). Ethics of interventions for the welfare of free-living wild animals. *Animal Welfare*, *5*(3), 235–244. https://doi.org/10.1017/S0962728600018820

Marion, J., Dvorak, R., & Manning, R. E. (2008). Wildlife feeding in parks: Methods for monitoring the effectiveness of educational interventions and wildlife food attraction behaviors. *Human Dimensions of Wildlife*, *13*(6), 429–442. https://doi.org/10.1080/10871200802270158

McCumber, A., & King, Z. (2020). The wild in fire: Human aid to wildlife in the disasters of the anthropocene. *Environmental Values*, *29*(1), 47–66. https://doi.org/10.3197/096327119X15579936382671

Michelfelder, D. P. (2018). Urban wildlife ethics: Beyond 'parallel planes'. *Environmental Ethics*, *40*(2), 101–117. https://doi.org/10.5840/enviroethics201840212

Monath, T. P. (2013). Vaccines against diseases transmitted from animals to humans: A one health paradigm. *Vaccine*, *31*(46), 5321–5338. https://doi.org/10.1016/j.vaccine.2013.09.029

Morris, M. C., & Thornhill, R. H. (2006). Animal liberationist responses to non-anthropogenic animal suffering. *Worldviews*, *10*(3), 355–379. https://doi.org/10.1163/156853506778942077

Narveson, J. (1977). Animal rights. *Canadian Journal of Philosophy*, *7*(1), 161–178. https://doi.org/10.1080/00455091.1977.10716186

Ng, Y. -K. (1995). Towards welfare biology: Evolutionary economics of animal consciousness and suffering. *Biology & Philosophy*, *10*(3), 255–285. https://doi.org/10.1007/BF00852469

Palmer, C. (2021). Assisting wild animals vulnerable to climate change: Why ethical strategies diverge. *Journal of Applied Philosophy*, *38*(2), 179–195. https://doi.org/10.1111/japp.12358

Patton, M. L., Jöchle, W., & Penfold, L. M. (2007). Review of contraception in ungulate species. *Zoo Biology*, *26*(4), 311–326. https://doi.org/10.1002/zoo.20154

Ray, G. (2017). How many wild animals are there? *Wild Animal Suffering Research*. Retrieved June 2021, from https://was-research.org/writing-by-others/many-wild-animals/in

Rupprecht, C. E., Hanlon, C. A., & Slate, D. (2004). Oral vaccination of wildlife against rabies: Opportunities and challenges in prevention and control. *Developments in Biologicals*, *119*, 173–184.

Sagoff, M. (1984). Animal liberation and environmental ethics: Bad marriage, quick divorce. *Osgoode Hall Law Journal*, *4*(2), 297–307. https://doi.org/10.13021/G8PPPQ.41984.1177

Sapontzis, S. F. (1987). *Morals, reason, and animals*. Temple University Press.

Sássi, C., Carvalho, G. D., de Castro, L. M., Junior, C. Z., Nunes, V. D. F. P., Do Nascimento, A. A. T., Bastos, A. L. F., Santana, L. G., & Do Nascimento Gonçalves, I. (2021). One decade of environmental disasters in Brazil: The action of veterinary rescue teams. *Frontiers in Public Health*. Retrieved June 2021, from https://www.frontiersin.org/articles/10.3389/fpubh.2021.624975/full/in

Sawyer, J., & Huertas, G. (2018). *Animal management and welfare in natural disasters*. Routledge.

Sebo, J. (2022). *Saving animals, saving ourselves: Why animals matter for pandemics, climate change, and other catastrophes*. Oxford University Press.

Soryl, A. A., Moore, A. J., Seddon, P. J., & King, M. R. (2021). The case for welfare biology. *Journal of Agricultural & Environmental Ethics*, *34*(2), 1–25. https://doi.org/10.1007/s10806-021-09855-2

Thompson, K., Leighton, M. A., & Riley, C. (2015). Helping hands, hurting hooves: Towards a multidisciplinary paradigm of large animal rescue. *Australian Journal of Emergency Management*, *30*(2), 53–58.

Tomasik, B. (2009/2019). How many wild animals are there? *Essays on Reducing Suffering*. Retrieved June 2021, from http://reducing-suffering.org/how-many-wild-animals-are-there

Tomasik, B. (2010/2015). The importance of wild animal suffering. *Relations: Beyond Anthropocentrism*, *3*(3.2), 133–152. https://doi.org/10.7358/rela-2015-002-toma

Torres, M. (2015). The case for intervention in nature on behalf of animals: A critical review of the main arguments against intervention. *Relations: Beyond Anthropocentrism*, *3*(3.1), 33–49. https://doi.org/10.7358/rela-2015-001-torr

Turnbull, P. C. B., Tindall, B. W., Coetzee, J. D., Conradie, C. M., Bull, R. L., Lindeque, P. M., & Huebschle, O. J. B. (2004). Vaccine-induced protection against anthrax in cheetah (*Acinonyx jubatus*) and black rhinoceros (*Diceros bicornis*). *Vaccine*, *22*(25–26), 3340–3347. https://doi.org/10.1016/j.vaccine.2004.02.037

Unger, P. (1996). *Living high and letting die*. Oxford University Press.

OPEN ACCESS

Welcoming, Wild Animals, and Obligations to Assist

Josh Milburn

ABSTRACT

What we could call 'relational non-interventionism' holds that we have no general obligation to alleviate animal suffering, and that we do not typically have special obligations to alleviate wild animals' suffering. Therefore, we do not generally have a duty to intervene in nature to alleviate wild animal suffering. However, there are a range of relationships that we may have with wild animals that do generate special obligations to aid – and the consequences of these obligations can be surprising. In this paper, I argue that we have special obligations to those animals we have historically welcomed or encouraged into our spaces. This includes many wild animals. One of the consequences of this is that we may sometimes possess obligations to actively prevent rewilding – or even to *de*wild – for the sake of welcomed animals who thrive in human-controlled spaces.

Over the past decade, a philosophical case that we have strong obligations to aid wild animals in need has emerged (see, e.g. Faria, 2023; Horta, 2017; Johannsen, 2021). After all, it is argued, there is a great deal of suffering in nature, animals' suffering matters, and we could be doing something to alleviate that suffering. So, we should.

This is far from a solely theoretical endeavour. Activist organizations have begun to support providing aid to wild animals. This is not to conserve species or ecosystems, nor to protect wild animals from human encroachment. It is to help wild animals overcome the challenges they face in the natural course of their lives. At time of writing, Wild Animal Initiative is one of Animal Charity Evaluators's 'Top' charities (ACE, 2022). Part of the reason for this is the relative neglect of wild animals among activists, but its inclusion in the list nonetheless indicates that wild animal suffering is far from a niche concern in some advocacy circles.

Nonetheless, lots of animal ethicists and activists are surprised by the conclusion that we should be intervening to alleviate the suffering of wild animals.[1] If we want to resist the conclusion, there are several strategies we could adopt. We could argue that wild animals may have relatively good lives (cf. Browning & Veit, 2023), that harm to animals should not motivate us to act (cf. Ebert & Machan, 2012, pp. 152–154),[2] or that the value of natural relationships outweighs harm to animals (cf. Sagoff, 1984). Or, we could acknowledge the *theoretical* viability of intervention, but challenge the feasibility, practicality, or utility of championing intervention (cf. Delon & Purves, 2018).

Some of the most interesting challenges to intervention, however, have come from people who affirm the value of animals, yet offer principled arguments against intervention. They affirm the significance of the different relationships that humans have with animals, arguing that the relationships that humans have with wild animals justify (or demand) a hands-off policy – we should, in the words of Tom Regan, 'let them be' (Regan, 2004, p. xxxvii). Champions of this approach include Clare Palmer (2010) in ethics, and Sue Donaldson and Will Kymlicka (2011) in political theory.

In this paper, I begin with this 'relational non-interventionism', and explore where it might take us. Strikingly, there are a range of possible relationships that people can have with wild animals that – even on this apparently non-interventionist account – justify strong positive duties. My focus will be on one that has been overlooked. Many wild animals have been historically encouraged and welcomed into spaces by humans. And this creates a relationship between humans and those animals that could ground strong duties to aid. These duties may have surprising consequences, even though they arise from an apparently conservative theoretical approach. Contrary to the claims of many environmentalists, who champion *re*wilding, we may have a *prima facie* duty to *de*wild certain spaces – or keep them from being rewilded. That is, we may have a duty to create or maintain artificial, apparently human-dominated spaces for the sake of wild animals. Among the many and various proposals for what taking animals seriously might require of us, this is one that has gone unexplored.

Note, incidentally, that this is not a proposal simply arguing that dewilded spaces will involve less animal suffering than wild or rewilded spaces, so we have a duty to dewild. So-called *pavers* argue that we should 'pav[e] paradise and put up a parking lot', as sung by Jodi Michell. The argument is that

> we have very good reason to believe that destroying habitat will reduce the number of wild animals who come into existence, and if most wild animals have net negative lives, then reducing the number of wild animals who come into existence will increase total utility. So, insofar as we ought to increase total utility, we have moral reason to destroy animals' habitats. (Fischer, 2022, p. 865)

I am no paver. Nor am I arguing that we should take control of habitats and ecosystems to minimize suffering, such as by favouring ecosystems replete with large herbivores (Horta, 2017, p. 275). That may be a kind of dewilding. Instead, I argue that the obligations we have to *particular* wild animals sometimes give us a reason to create or preserve non-wild spaces. This argument is independent of any claim about minimizing suffering.

The paper will advance as follows. I will begin by outlining the theoretical background of the argument. I will then defend my core claim: That A's historical welcoming or encouragement of B grounds certain key *prima facie* positive duties towards B for A. Examples follow, illustrating how this innocuous-sounding duty provides support for a surprising conclusion: We are sometimes obliged to dewild spaces for the sake of animals who do (or would) live there.

Interventionism and Non-Interventionism

Many defences of intervention to limit the suffering of wild animals – hereafter, *interventionism* – are consequentialist in character (see, e.g. Horta, 2017; Sebo, 2022). It is easy to see why. There is much suffering in nature. We could lessen that suffering. And if we are consequentialists[3] we will believe, roughly, that if we can do something to lessen suffering, then, all else equal, we should. The devil is in the details – for example, it might not be clear that we *could* do something to lessen this suffering. But even if a particular consequentialist is a non-interventionist in practice, they should support intervention in principle. Perhaps, indeed, there is good reason to think that they should today support tentative first steps towards future intervention (see Ng, 1995). Examples of these first steps might include research on ecology, genetic editing, political practicalities, veterinary epidemiology, and so forth.

Other non-interventionists are non-interventionist in principle, as well as in practice. That is, they oppose intervention (or permit non-intervention) not because of worries about feasibility, effectiveness, or similar, but because they do not think that suffering alone is a sufficient reason for intervention.

Some principled defences of non-interventionism can be put aside. For example, we could think that the suffering of animals is not a particularly compelling reason to act, and reject interventionism for that reason (cf. Ebert & Machan, 2012, pp. 152–154). Some others might affirm that the value of 'the natural' (or similar) is sufficient to give us decisive reason *not* to intervene in nature (cf. Sagoff, 1984). Other defences of non-interventionism, however, do not seek to either diminish the significance of animal suffering, or to argue that non-intervenion is justified by some other value. It is these defences with which I am presently concerned.

The argument goes like this. All animal suffering matters. But the extent to which a particular instance of animal suffering grounds a duty for a particular agent to act depends upon the relationship that said agent has to the suffering or the animal experiencing it. To be obligated (or even permitted) to aid, the agent must have a particular kind of normatively salient relationship to the suffering or animal. To put it another way: There is no *general* obligation to aid animals in need, but facts about our relationships with particular animals may generate *special* obligations to aid them when they are in need.

This may sound abstract, but it is perfectly familiar. My failure to aid my dog when she is suffering because I have chained her in a barren yard without food or water seems monstrous. My failure to aid a stray dog who frequents my street when she is suffering because she is without food or water may seem heartless, but it is hardly monstrous. My failure to aid a wolf I see upon a hike who is clearly starving is unlikely to evoke much moral disapproval at all. In each case – beloved companion, homeless stray, wild wolf – the suffering is comparable, and the suffering beings themselves are comparable. (There are not many differences between companion dogs, stray dogs, and wild wolves – all are sentient, all have similar intellectual and emotional complexity, and so forth.) But the relationship between the suffering (or the suffering being) and me are different in each case, and thus the special obligations I possess are different.

In the case of the companion, there are a host of relevant relationships, and all could be special-obligation-generating. The companion's suffering has been caused by me. The companion's dependence on me has been caused by me. The companion's *existence* may have been caused by me. There are likely to be certain affective relationships (love, friendship, admiration) between me and the companion – or there may once have been. And so on.

When it comes to the wolf, there are no such relationships. I have no straightforward causal relationship with the suffering of the wolf, and nor do I have any straightforward causal relationship with the wolf herself. (That is, and perhaps in sharp contrast to my companion dog, I am not straightforwardly responsible for her existence, the vulnerabilities/dependencies fostered by her genetics,[4] or similar.) I have no affective relationship beyond the one created by my seeing her at this instant. Thus, it is possible that I have, at most, very few special obligations concerning this wolf.

The other case falls somewhere in the middle. I have *some* relationship with the stray – I know her as a neighbour of sorts, and perhaps recognize that I am part of a community that has led to her being in need. These may generate weak special obligations. Perhaps, then, we might think the following: I am *required* to aid my dog, and it would be *unjust* for me not to do so. It would be *good* for me to feed the stray, and *bad* for me not to. It would be *supererogatory* for me to feed the wolf, but it would *not be wrong* for me to fail to do so.

I do not commit to that conclusion[5] – my point is that such a conclusion makes sense. But it makes sense even though the level of suffering and facts about the suffering being herself are comparable in each case. An argument like this gives us a non-consequentialist response to the interventionist. It gives us a non-interventionism that, first, is principled (i.e. it is not opposed to interventionism in practice only), and, second, affirms the high value of animals.

But perhaps this is too quick. Why deny that we have general obligations to aid wild animals? Incidentally, one does not need to be a consequentialist to recognize such obligations. Perhaps we have a duty of beneficence to aid wild animals in need (Johannsen, 2021). Denying such a duty forces a dilemma upon the animal ethicist. It seems to entail that *either* there is no duty of beneficence to humans in need, *or* it entails that we have different kinds of obligations to humans and animals. The former horn of the dilemma sounds surprisingly libertarian – relatively few ethicists will embrace the conclusion that we have no obligations to aid distant human strangers in need.[6] (That said, the libertarian can endorse the claim that we have duties to aid distant strangers – they will just deny that they are duties of justice.) The latter sounds surprisingly speciesist – relatively few animal liberationists will embrace the idea that there is a bright moral line between humans and animals.[7]

Palmer (2010) defends relational non-interventionism. A close look at her response to the possibility of a duty of beneficence will be instructive. She borrows from Samuel Scheffler (1997) to argue that moral responsibilities can 'emerge out of membership in some common group' (Scheffler, 1997, p. 190), and from Leslie Pickering Francis and Richard Norman (1978) to identify the common group created by the 'network of rich interhuman relations' (Palmer, 2010, p. 121). The problem[8] is that this common group seems to include many animals. The 'global human community should, to be consistent, be extended to a global "mixed" social community, including all humans and the nonhuman sentient animals with whom humans have strong social relations' (Palmer, 2010, p. 123). Thus, the global mixed social community may include wild animals, undermining Palmer's principled non-interventionism: If duties of beneficence are owed to other humans because they belong to the global social community, but wild animals *also* belong to that community, duties of beneficence are owed to wild animals – and thus relational non-interventionism ends up advocating intervention.

The question becomes whether wild animals are a part of this mixed social community. Donaldson and Kymlicka (2011) stress the *particular* community formed by the co-relations between humans and domesticated animals (see also Smith, 2012; Valentini, 2013). Alasdair Cochrane (2018), meanwhile, argues that humans, domesticated animals, and wild animals *all* share in a 'community of fate', and thus that, to paraphrase Palmer, 'strong social bonds' are formed between humans and wild animals. How much Palmer's affirmation of a mixed human/animal social community undermines her support for non-interventionism depends on how large the social community is appropriately thought to be.

Bonds between (potential) members of a social community are not the only special-obligation-generating relationship that may be important in responding to relational non-interventionism. In many cases, humans have morally salient relationships with the harm experienced by wild animals, as they have *caused* the harm. A significant – albeit complicated – example is anthropogenic climate change (Palmer, 2021). Due to climate change, particular animals lose food and habitat; are impacted by extreme weather; are displaced or predated by animals fleeing other habitats; and so on. At the extreme, species are driven to extinction. Given the causal relationship between the harm experienced by these animals and humans, the animals have a claim to human assistance. The case is difficult (partly) because it is impossible to identify *particular* humans who are responsible. Questions about putting right the wrongs of climate change to humans, let alone animals, have prompted a complex and contentious academic literature.

Beka Jalagania (2021) explores another relationship that can maintain between humans and wild animals in need. Many humans *benefit from* wild animals. We have a responsibility to share benefits with those who contribute to their production, and we can share benefits with wild animals through minimizing their suffering. The examples that Jalagania draws upon include photographers benefitting from charismatic wildlife; stressed people benefitting from birdsong; and documentary-watchers benefitting from the wildlife profiled. But the argument could be expanded. Given that humans benefit from the 'ecosystem services' animals provide, we may have special obligations to assist many different wild animals (cf. Blattner, 2020).

These arguments – drawing upon social bonds with wild animals, causal responsibility for wild animals' plight, and benefitting from wild animals – show the following. Even if we accept the relational non-interventionist's claim that there is no *general* obligation to aid wild animals, there are a range of *special* obligations that may ground a duty to intervene.[9] Thus, the relational non-interventionist position turns out to be not nearly as non-interventionist as it first seemed.

My objective in the remainder of the paper is to draw attention to another such relationship. This is a worthwhile activity for two reasons. First, it outlines another set of circumstances in which intervention to aid wild animals may be required, even for the relational non-interventionist. Second, this special-obligation-grounding relationship has some surprising consequences. In particular, it sometimes offers a justification for what we might call *de*wilding: The creation or maintenance of artificial spaces, for animals' benefit. More on that later.

Historical Encouragement

My proposal is that one special-obligation-grounding relationship between humans and some wild animals is that the latter have been *historically welcomed* by humans. By this, I mean that these wild animals have been deliberately encouraged to inhabit certain areas or behave in certain ways; I do not simply mean that humans have created a space that happens to be attractive to them.

If I invite someone into my home, I gain some responsibility for their wellbeing. If I open my garden to the public, I gain some responsibility for visitors' wellbeing. If I build a park for local children, I have some responsibility for the wellbeing of the children while they play. If my house is decrepit, my garden overrun with biting insects, or my park prone to fall apart, visitors have some claim against me. I would not have the same level of responsibility if people just started visiting my house, garden, or park without my knowledge or blessing, *even if* there is a clear sense in which I share in some *causal* responsibility for any injuries that befall the visitors. (If I did not have an enticing home, garden, or park, no one would have entered.)

But there is also a stronger sense in which individuals might have a claim against someone who has encouraged their presence. Imagine a landowner creates a campsite for people to live in, giving every impression that this space will be theirs to live in indefinitely into the future. She does not cede ownership rights or anything so extreme, but she encourages people to join her in this space, gives no sign that the space will change uses, and welcomes them there with things they like – food, amenities, protection. Imagine that the landowner is so welcoming that some of her guests 'sell up' and move in permanently, siting mobile homes and similar. The residents would have a claim against the landowner were she to declare one day that the site was going to become a lake. (Indeed, there is a chance they would have such a claim in law, and not 'merely' a moral claim – but legal rights are not my focus here.) This might be *in part* because the campers have become dependent on the landowner. In this case, that would include, but not be limited to, a dependence on her inaction. But it is deeper than this; it is because the landowner has *encouraged* them to be present at this site that there is a special obligation that she seems to be violating in turning the site into a lake.

To see why, consider the case of individuals who set up caravans without the landowner's blessing – without her knowledge, perhaps. In this case, the residents foster a dependency on the landowner; it is her land, they (let us say) take food from her stores, and so on. With or without permission, the residents are dependent upon the landowner, but in one case there seems to be a particular wrong in the landowner turning the site into a lake. This is because of the particular relationship she has with the residents and/or their current situation. Not only do the residents rely upon her in some sense, but she seems to be responsible for this, through her positive action.[10]

Might the wrong be explained in terms of 'legitimate expectations' (Brown, 2017)? The thought is that A's legitimate expectation about B's future conduct is something that should influence B's decision-making. So, in this case, the residents' legitimate expectation about the landowner's maintenance of the site should influence the landowner's decision about the lake. *Prima facie*, the residents would be wronged, as the landowner would act against their legitimate expectations, and thus disrupt their life in significant and morally questionable ways. But the residents need good grounds for their expectations about the site to be *legitimate* expectations. The landowner's welcoming provides this.

Another way to think about the landowner's wrong is in terms of trust – the residents trust that the landowner will maintain the site, and her wrong is violating that trust. But why do the residents trust the landowner? It is precisely because she has welcomed them. Trust, like legitimate expectations, seems to be another way to underline the moral significance of welcoming.

So, invitation, welcoming, and explicit encouragement seem to form a special-obligation-grounding relationship. I have left open the precise nature and strength of the special obligation created. For example, it is consistent with what I have argued that only a limited obligation is created. Take my landowner case: it is possible that the strength of the special obligation pales in comparison with the strength of the landowner's right to do what she will with her land.[11] Or perhaps the special obligation can be discharged relatively easily. In the case the open garden, perhaps I could discharge my obligation by putting up a warning sign or providing medicinal cream. The obligation being comparatively weak or easily discharged differs, however, from it not existing.

It is my contention that the duty I have outlined applies to many cases of interactions between wild animals and humans. That is, humans have encouraged the presence of particular wild animals, and thus have some responsibility for their wellbeing, and a *prima facie* duty to aid them. Not only is this consistent with the relational approaches associated with non-interventionism, but it has some surprising consequences.

Objections

Distinctiveness

Perhaps I have not identified a distinctive obligation. It may sound like I am not arguing that we have an obligation to aid certain others, but that we have an obligation not to harm certain others. This is not so. In the case of the creation of a lake, I am plausibly talking about something harmful. (My reason for this case will become clear later.) But in the case of the biting insects, I am talking about acting to prevent the harm of irritating bites (like a sign) or acting to alleviate it (like providing abirritants). While historical encouragement may generate special obligations not to harm that supplement existing duties not to harm,[12] it also generates special obligations to aid. For example, I have a *prima facie* duty to aid those invited into my garden when they are hurt – I do not have the same duty to aid those who hurt themselves in my garden when they have not been invited. If a passing pedestrian wanders across my lawn and twists her ankle, there is a sense in which it is not my problem. If, however, I invite the pedestrian to see my flower beds, and she twists her ankle, I have a *prima facie* duty to offer her aid: fetch her a chair, prepare a cold compress, or similar. While the proposed special obligation can manifest in duties of non-maleficence, it can also manifest in duties of beneficence.

It may also sound like I am simply reiterating the already-discussed special obligation generated when an agent is responsible for the harm experienced by another. In a limited sense, this is true. Saying that A encourages B to do an action that leads to harm to B is not the same thing as saying that A is causally or morally responsible for the harm that befalls B. But it is usually to say that A has *some* causal or moral responsibility for B's harm. If an invited visitor to my decrepit house falls down the stairs, I have some moral and causal responsibility for the harm of the fall, even if not as much as if I literally pushed her.[13] But these things exist on a continuum. It would be a stretch to say that I did not kill a mouse but merely 'encouraged' her into a trap by placing some cheese. Equally, though at the other end of the spectrum, it would be implausible to say that I killed a mouse by laying out cheese for her, if, on the 100th night I put out food, she is killed by an owl I had no reason to believe visited my garden.

Understanding

Is there any reason to believe that we would not have obligations of this sort to wild animals, even while we do with humans? One potentially relevant difference is that wild animals do not understand, we may think, what it means to be welcomed or invited. Many humans, on the other hand, do.

The first thing to say is that individual wild animals can and do interpret cues regarding invitations and welcomes – and, on the flip side, rejections and warnings. And this can lead to personal relationships between individual humans and individual wild animals developing. But I do not want to rest on the potential for clear communication between humans and animals to make sense of obligation I am proposing.

This leads to the second response. We can make sense of the idea that encouraging or welcoming individuals into our spaces generates obligations to those individuals regardless of whether the individuals understand the welcoming and/or encouragement. If I am socializing with a mixed-species group of humans and dogs, I do not need to worry much about whether the dogs *understand* or *appreciate* my invitation when I place a collection of toys on a lawn. I am doing something wrong if I place the toys in a place that will be harmful for them – say, if the lawn has been treated with aggressive pesticides – precisely because, in welcoming them, I am assuming some level of responsibility for their safety. (The degree to which we take on this responsibility depends on a multitude of factors.)

One way of putting this is the language of *hospitality*. When one welcomes another into their space, one takes 'responsibility for the overall welfare' of the guest. 'Traditionally the most important responsibility of all was for the guest's safety – hospitality was a kind of sanctuary, and the host was thought of as having undertaken a solemn obligation to make sure no harm came to his guest while under his roof' (Telfer, 1996, p. 83). But nothing about the 'solemn obligation' of hospitality depends upon the object of hospitality understanding herself to be an object of hospitality.

It might sound odd to talk about extending hospitality to animals, but this is not so. Hospitality is 'the giving of food, drink, and sometimes accommodation to people who are not regular members of a household' (Telfer, 1996, p. 83). If we allow a sufficiently flexible account of 'people', then the welcoming of animals into our spaces – say, feeding garden birds – seems like a paradigm example of hospitality (Milburn, 2022, chap. 4). Indeed, a close look at hospitality tells us much of interest about the present case. Hospitality carries with it a certain kind of intimacy (Telfer, 1996, p. 93). The extension of hospitality is not about company in general – it is about company with the particular other to whom hospitality is extended. And so it carries with it particular obligations to that other, and not just to anyone. Indeed, we are not obliged to extend hospitality to everyone. Hospitableness is, at best, an imperfect duty, or optional virtue (Telfer, 1996, p. 96); it cannot be offered to everyone. It may be right to refuse to extend it in some cases

(Boisvert, 2014, p. 1184). For instance, my duties of hospitality towards the lamb prevent me from extending hospitality to the wolf (Milburn, 2022, chap. 4).

But, might say my critic, if an idea of trust is doing work in the examples I am forwarding, animal understanding does seem to be presupposed. In important senses, it *does* seem like animals can trust us, that we can violate that trust, and that this violation entails important wrongs (Cooke, 2019). So, it is plausible that a trust-based account could be applied to animals, including wild animals.

Something similar could be true of legitimate expectations. Does *this* idea not presuppose understanding? It is the legitimacy of the expectation, rather than the expectation itself, that grounds the normative salience of the legitimate expectation – it is the legitimacy of the expectation that means that one has been *prima facie* wronged when the metaphorical rug is pulled from under one's feet. The wrong comes from the fact that someone has planned around the expectation – up to and including ordering their life around the expectation (think of the campers who 'sell up'). Yet, through no fault of their own, this expectation is shattered. Let us be clear that, just as humans can, animals can structure their lives around expectations that they will continue to be able to utilize a certain space. And let us observe, once again, that to welcome another – human or animal – to a space is to offer them some assurance that they may safely use that space. It seems, therefore, that the idea of legitimate expectations can make sense of both the special obligation that intuitively exists in the cases explored, and the importance of the welcoming as an explanatory factor – even without positing full understanding on the animals' part.

Those wanting to argue that animals' lack of understanding undermines my argument may find that their objection backfires. It is conceivable that animals' lack of understanding means that we are *more* likely to have obligations towards them. This is a possibility offered by Gary O'Brien (personal communication, 14 July 2021). He allows that inviting people into a space grounds duties in a way that merely enticing them (say, by cultivating a particularly attractive space) does not. However, when it comes to animals, it is

> plausible that I have strong duties towards them even if I merely enticed, but didn't invite, them in. This is because (i) animals aren't moral agents – they can't be held responsible for 'trespassing' on my property in the way that adult humans can and (ii) animals are generally less competent to navigate dangers (especially unfamiliar manmade dangers) than adult humans are, and hence they are more vulnerable. (personal communication, 14 July 2021)

This means that, assuming I have built a very attractive playground in my back garden, I am not morally responsible for injuries experienced by adults who invite themselves in. But children lack moral agency, have lower capacities to resist, and are more vulnerable to harm than most adults. Thus, if a child enters – thinks O'Brien – I may be morally responsible for any harm she faces. Many animals are relevantly like children in these regards: that is, they lack moral agency, they have lower capacities to resist temptation than do adult humans, and could be more vulnerable to harm than adult humans. (Of course, this is not to deny that there may be many *other* morally salient differences between human children and animals, and it is not to say that all animals are alike when it comes to, say, their vulnerability.) If this train of thought is heading in the right direction, then one may gain obligations towards children, animals, and other relevantly similar individuals merely through enticing, even if not actively inviting, them. The line between (say) the birds I have *invited* through hanging feeders and the mice I have *enticed* through planting beans is blurred, morally speaking. But let us put this possibility to one side, and remain with the idea that it is inviting, and not merely enticing, that is special-obligation-grounding.[14]

Countervailing Considerations

Might we have other reasons to be sceptical about extending this duty to wild animals? Proposals to aid wild animals are frequently met with scepticism because of humans' poor track record when it comes to intervening in nature. Let us concede that human interventions in nature have often had very bad results for animals. Donaldson and Kymlicka point to a case of mass-culling, supposedly in the interests of conservation, when the choice to cull was based upon a model of populations that did not account for natural patterns of population growth and decline (2011, p. 163). In this case, humans' limited understanding of ecology had bad results both for animals and for conservation goals.

This does not undermine my argument. It is possible that the wrong of imposing risks on animals through ham-fisted interventions outweighs the *prima facie* duty to aid generated by our relationships when our understanding is limited. But this is not to deny that there is a *prima facie* duty to aid. It is simply to argue that, in some cases, it would be inappropriate to act upon the *prima facie* duty, as there are weighty countervailing concerns. Indeed, it might suggest that we have reason to actively improve our understanding, so that we are in a better position to aid in the future.

Does this suggest that the proposed special obligation is impotent? It is one obligation among many. And its strength and reach will vary. Putting out seeds for birds in our garden does not oblige us to care for those birds for the rest of their lives, but it does carry with it obligations to make sure that our spaces are minimally bird-friendly. For example, it would be wrong to put out seed when we know that there are multiple bird-hungry cats in the garden. It would even be wrong – though perhaps excusable, given people's ignorance – to put out bird seed when (because of dirty feeders) doing so will spread disease.

Garden birds – typically – have plenty of other options than eating in *this* garden. Small obligations are generated by one-off or intermittent welcomes. Weightier obligations are generated by welcoming someone indefinitely. Weightier obligations still are generated by welcoming someone when said welcoming leads to them having few or no other options. Indeed, there may be a fine line between indefinite welcoming of a wild animal and domestication. Intuitively, we have small obligations to an occasional houseguest while she is a houseguest. We have greater obligations to someone who we have invited to live in our house. And we have greater obligations still concerning the person who, in accepting our invitation to come and live in our house, burns metaphorical bridges. Short of serious countervailing considerations, it would be monstrous for (to retell the Parable[15] for present purposes) a Samaritan to welcome a man into his home, knowing that this means that the man now faces indefinite ostracism from his people, only to cast him out again a few days later. What is striking, however, is that this precisely what we do with some wild animals.

There is another complicating factor, too. It is one thing to talk about the welcoming of a particular discrete individual by a particular discrete agent, and this agent's consequent obligations towards the invitee. But what is often going on in the case of wild animals is that *groups* of wild animals have been invited and/or welcomed by *groups* of humans – sometimes recently, sometimes historically. While considerations like this certainly seem to *lessen* the obligations owed to particular wild animals (and make identifying the agents responsible for these duties difficult), it is surely plausible that the invitation and/or welcoming continues to carry normative weight.[16]

Indeed, it is *precisely* this sort of historical, population-level interaction that motivates the relational non-interventionist's claims that we have duties to domesticated animals that we do not have to wild animals. In an important passage (cf. Sachs, 2019, pp. 94–95), Donaldson and Kymlicka (2011, pp. 79–80) compare the history of domestication to the North Atlantic slave trade, arguing that even if domestication was unjust, it grounds continuing positive obligations to domesticated animals.

Case Studies

The special obligation defended in this paper could have some surprising results. The fact is that we have *not* encouraged or welcomed all wild animals. But we have encouraged some – and we have encouraged them into very particular spaces. What this means is that we might have an obligation not to 'rewild' these spaces, even when the opportunity arises. Instead, we have at least *prima facie* reason to *de*wild them – to preserve them as, or make them into, or return them to, human-controlled spaces, *for the sake of the animals who live there*.

In this section, I will offer two case studies displaying the starkness of this proposal, demonstrating what a relational approach could mean for duties to wild animals. These animals are not paradigmatic wild animals – Donaldson and Kymlicka would frame them as *liminal* animals – but they live outside of human control. They are free to come and go; they are 'unowned'. They are not *domesticated* in the Latin sense – that is, they are rarely present in houses.

Urban Animals

Historically, people have welcomed and encouraged certain wild animals in urban environments. This continues today. Ponds are dug in parks to attract ducks, where they are fed by children. Birds of prey take up residence in boxes on towers and cathedrals, where they are monitored to the delight of naturalists. Cats and dogs roam streets, sometimes fed, sometimes befriended, sometimes sheltered. What is clear is that some wild animals are welcomed into urban environments; their presence is encouraged. As argued above, this creates a particular kind of relationship between humans and these animals, grounding a *prima facie* duty to aid the animals when they are facing hardship.

One kind of hardship faced by these animals comes about when their relationship with humans is artificially severed – if, for example, new bylaws prohibit the feeding of pigeons in places where they were previously encouraged (Palmer, 2003). Now, this is not to say that we necessarily wrong pigeons with such bylaws. But it is to say that we owe them consideration when these laws are being crafted – alternatives should be offered. If this a matter of a few pigeons no longer being fed in a particular location (say, a hospital deciding to no longer hang feeders in a courtyard), then that is one thing. The pigeons can find food elsewhere, and so are not in particular need of aid. But if it is a case of many pigeons, or pigeons over a large area, no longer being fed, the duty grounded by the historical relationship calls for more attention, precisely because these pigeons may now find themselves in need. Could the feeding be phased out gradually? Could the pigeons be welcomed elsewhere? Could contraception humanely control pigeon populations to ensure minimal harm? I am not saying that these options are uncontroversial – but they are the sorts of things that need to be considered, precisely because of the historical welcoming.

In contrast, those urban animals who have *not* been historically welcomed – rats, say[17] – need not be considered to the same extent. Consider a change in food-waste disposal practices, leading to rats having less access to food. This may adversely impact rats, but the different kind of relationship we have with them provides at least *a* reason to say that we do not need to be *as* concerned as we are obliged to be about the pigeons. Or consider gulls, who often thrive at landfill sites, feeding on scraps. They are not welcomed at these sites, and so we should not be concerned about our impact on the gulls if we change our landfill practices. Or if we should, it is not because we have welcomed them.

Of course, many animals are welcomed by some humans, and actively discouraged by others. While urban pigeons (descendants of domesticated birds, complicating matters further) are fed by tourists and cared for by activists, they also face hostile architecture designed to keep them off buildings and bird feeders designed to keep them out. Are these animals welcomed, in the morally salient sense? Surely, the fact they have been welcomed by *some* people gives them a claim to assistance, even if not as much as more universally welcomed animals. Responsibilities, too, are going to lie more strongly with those involved in welcoming than with those who have not welcomed them – though this may not be to deny that we all have *some* responsibility towards urban pigeons that we do not have towards wild rock doves. There is a messiness, here. But that reflects the complicated and paradoxical relationships and history that we have with these animals.

But changes impacting welcomed urban wildlife could be much broader than simply a change in bylaws around feeding. Let us envision a particularly stark possibility. Imagine we saw massive, rapid human depopulation, in response to a pandemic, changing birth trends, or similar. Imagine this rapid decline in population saw some cities almost entirely depopulated. This would be a bad thing for the urban wildlife depending upon our presence in cities, including those we have welcomed – who may have come to rely on the warmth, light, shelter, and especially food generated by our presence.

Normally, the idea of depopulated cities raises the prospect of *rewilding* spaces – making these spaces 'natural', allowing animals to live free from human interference. But what I am suggesting is that, given the special obligations we have to certain urban animals, we have at least some reason to be opposed to rewilding these spaces. Instead, we have *prima facie* reason to continue to 'use' these spaces: to continue to put up nest boxes, spread food, to heat buildings in the heart of winter, and prop up collapsing buildings. In a word, there is good reason to think that we should *dewild* these spaces, *for the sake of the animals who inhabit them*. After all, they benefit from having access to these non-wild spaces, and might not have any other (potential) home – and we have special obligations, because we have historically welcomed them.

Farmland Animals

One might object that many of the animals who thrive in urban environments – invited or otherwise – are highly adaptable. Indeed, that is the point; they have been able to move from the 'natural' environments favoured by their ancestors to develop new ways of life in cities.[18] Even if humans were to stop using these urban spaces, these animals – it may be thought – would move on to new environments where they could thrive.[19] But some animals living in human-controlled spaces are specialized, meaning that changes in human practices leave them in danger. Take farmland birds. These are animals who have traditionally thrived in ecological niches created by human farming practices. Changing agricultural methods – such as changing use of pesticides and other chemical interventions, more intensive land use, mechanization, and so on – present threats to individuals and whole populations. In the United Kingdom, for example, farmland specialist species such as corncrakes and gray partridges underwent dramatic declines in the latter half of the 20th century, moving from familiar rural birds to localized populations under threat of extinction.

In response to population decline, some conservationists, landowners, and farmers have sought to change farming practices to explicitly encourage the animals to come back. So, for example, certain farmers have left larger areas uncultivated, or even simply changed how they utilize farming machinery.[20] Now, actions like this could be taken to simply prevent adverse effects on wildlife. That would surely be commendable. But when it comes to the 'conservation' of charismatic birds like partridges, buntings, and larks, actions stretch beyond non-maleficence to an explicit encouragement of the animals' presence and a welcoming if/when they arrive. Actively encouraging the presence of these animals, as argued, generates special obligations towards them.

This historical encouragement of farmland birds gives us *prima facie* reason to support dewilding of a different kind. Imagine if we were to collectively move away from agriculture as it is currently practised, and towards indoor, vertical agriculture. This term refers to a range of agricultural practices from space-saving techniques used by home growers to proposals to grow food in skyscrapers. At its core is the idea that more efficient use of growing space can be made by growing *up*. In time, suggest indoor agriculture's optimistic proponents, all of the plant food needed by humans could be grown efficiently, safely, and reliably indoors.

There are a range of reasons that we might support indoor vertical agriculture, tied to concerns about food security, food sovereignty, food safety, climate change, and more. But it is worth mentioning that indoor agriculture has been occasionally championed because it could limit harms to wild animals (e.g. Lamey, 2019, p. 89; Milburn, 2022, chap. 5). Agriculture, even organic arable agriculture, causes harm to wild animals, such as those who are killed by farming machinery. A move to indoor arable agriculture, therefore, has the potential to limit harms to wild animals. One of the other putative advantages of indoor agriculture is that it allows space for rewilding. Once people are fed by space-saving indoor agriculture, agricultural land can be rewilded, with farmers retrained as caretakers (Platt, 2007).

But if we have special obligations to the animals encouraged to live in outdoor farmland – to corncrakes, buntings, partridges, larks, curlews, and the rest – this gives us a good reason to *not* rewild this land. (Of course, such a reason must be balanced against reasons we might have to rewild.) The argument against rewilding is bolstered by the fact that many of these historically welcomed animals are niche specialists. Unlike, it might be thought, the pigeons, foxes, peregrines, and so forth who have been encouraged and welcomed in cities, the farmland specialists need something very like *this* space. To rewild their spaces seems to be a wrong to them unless it is rewilded to a space suitable to them. This might entail *dewilding*, insofar as it is farmland to which they are, it seems, suited. At the very least, if *this* space is going to be rewilded, duties to those animals who have been encouraged could be discharged by the dewilding of some *other* space for their benefit. Maybe the land currently utilized as conservation-friendly farmland could be rewilded, while some other space could be dewilded, becoming an animal-friendly 'working' arable 'farm'.

Why the scarequotes around 'working' and 'farm'? I am not trying to argue against indoor agriculture, of which I am an advocate (Milburn, 2022, chap. 5). The reversal or prevention of rewilding – or even the complete dewilding of currently 'wild' spaces – for the sake of farmland animals is compatible with indoor agriculture. Farmers could continue to grow particular crops and manage the land, but primarily for the benefit of 'farmland' birds. Though this land has the appearance of a working farm, it is really a kind of simulation: A dewilded space for animals' benefit.

This will cost money, but that should not be a problem. First, any *re*wilding activity requires funding – and the proposals of indoor agriculture's proponents are no exception. If some of that rewilding money should be allocated to dewilding, so be it. Second, agriculture, and especially wildlife-conservation-friendly agriculture, is already subsidized. Again, if some of that money should be reallocated to dewilding, that need not be a negative. Third, collectively, we are already willing to fund schemes to help farmland birds. While dewilding might sound like a curious way to spend conservation money, that will mostly be a matter of framing. The difference between a 'rewilded' farmland and 'dewilded' nature is surely one of degree.

Conclusion: A Case for Dewilding?

Not only can a relational non-interventionist approach ground strong duties towards certain wild animals – thus challenging interventionism – it can ground very surprising duties. Namely, it can ground duties to dewild, here understood as the active prevention of rewilding, the active maintenance of non-wild spaces for the benefit of the animals who live there, and, at the extreme, the turning of 'natural' spaces into human-controlled, 'unnatural' spaces.

My focus here has been on the creation/maintenance of certain kinds of habitats for animals' benefit. These are environments in which food, water, shelter, and more are provided for the animals to access if they so choose. But a separate conversation needs to be had about the extent to which historical encouragement justifies, requires, supports (etc.) more hands-on forms of aid: vaccination and medical intervention; the 'adoption' of orphaned young; intervention in food cycles; genetic editing; and so on (cf. Johannsen, 2021). Further, though my case studies focused on large-scale, population-level relationships of encouragement, it should be clear that many of them can be adapted to an individual level. To repeat, nothing I have said here implies that anyone who feeds birds in their garden is obliged to care for these birds indefinitely into the future – not least because these birds surely have other options open to them. But perhaps people owe a little more to 'their' birds than they might to others.

Questions remain to be answered, especially about the strength of the obligations generated by (historical) welcoming. For example, I concede that I have conflated two different arguments. In one, the animals formerly encouraged to live in the land are the ancestors of the animals who currently live on the land. In the other, some of the animals formerly encouraged to live on the land are the *actual* animals who today live on the land. I think the obligation is clearly greater in the latter case. But there is plausibly a considerable obligation in the former case, too. After all, due to this historical welcoming, these animals may have nowhere else to go (Donaldson & Kymlicka, 2011, p. 79), and may be vulnerable and dependent because of human action (Palmer, 2010, p. 92). Indeed, these animals arguably exist because of us, which may be obligation-generating (O'Brien, 2021). But if this historical welcoming carries less weight, how much less? Just how beholden to our ancestors' welcoming of wild animals should we be?

I have also passed over complicated questions about the responsibility we might bear for the harm caused by the animals we have welcomed. Indeed, maybe the welcoming of predatory animals carries with it obligations to aid those prey animals they would otherwise harm (cf. Milburn, 2022, passim). More broadly, this opens contentious questions about conflicts between the interests of wild animals. What if helping the animals we have historically encouraged will be bad for *other* animals? I cannot offer formulas for working out what we should do in these cases. But I can point out that the duty I have defended here is a *prima facie* duty. It is entirely possible that other considerations will outweigh it in many real-world cases. But to work out whether they do, we need a fuller theory of our duties towards wild animals than I can offer here.

There is also further work to be done about the different kinds of encouragement and welcoming that humans have engaged in. For example, a particularly stark form of invitation and welcoming is the (deliberate) *introduction*. However, introductions are frequently disastrous. The ecological damage caused by cane toads and small Indian mongooses, for example, are well known. The treatment of 'invasive' species like these is a typical point of contention dividing animal and environmental advocates. But if, as I have argued, historical welcoming grounds a special obligation to aid welcomed individuals, it is possible that a double wrong is being committed against these animals if they are killed. Not only does the killing itself raise serious questions, but these are animals to whom, it seems, we owe particular *prima facie* duties of aid. It is my suggestion that we have at least some reason to hope that, in the future, rather than exterminating 'invasives', we will recognize that we have particular obligations to *aid* those 'invasives' here at our invitation.

The starkest contribution of the present paper has been to recognize that we will sometimes have reason to dewild spaces, specifically for the benefit of animals. This is not because we need more space to look after or support animals directly under human care. Nor is it an attempt to reduce suffering among wild animals by reducing the number of wild animals. Instead, it is because dewilded spaces are exactly where certain wild animals to whom we owe special obligations will thrive. In conversations about our duties to wild animals, this possibility has been overlooked. That, perhaps, is regrettable.

Notes

1. Indeed, critics have traditionally used it as something of a *reductio* of pro-animal positions (see, e.g. Sagoff, 1984).
2. Rainer Ebert and Tibor Machan do not commit to the claim that we have no duties to animals, though Machan (e.g. 2004) has defended something like this.

3. I am imagining utilitarianism, or a close relative.
4. Compare Palmer (2010, p. 92) on 'internal dependencies'.
5. Maybe there are other sensible readings. Perhaps I am required *not* to the feed the wolf – in so doing, I may illegitimately interfere with nature, or put other humans at risk.
6. We could adopt a libertarian approach to animal rights (Milburn, 2018).
7. Zamir (2007) is one self-confessed speciesist animal liberationist.
8. It is a problem both for non-interventionists and speciesists.
9. *May* because the special-obligation-generating relationships may not pertain, and because we may want to challenge the claim that a particular relationship *is* special-obligation generating.
10. Is being responsible for dependence the same thing as inviting someone? No – there are cases in which they come apart. Both, it seems, are plausibly special-obligation-generating relationships.
11. *This* is consistent with the claim that, though permissible, it would be morally suboptimal for her to turn the campsite into a lake.
12. Or create new ways in which we can harm. Perhaps replacing the campsite with a lake harms in the case in which the landowner has encouraged the presence of the campers, but not in the case in which she has not – in the latter, the campers have no right to the space. Perhaps we cannot be harmed by losing access to that to which we have no right. The mugger harms by taking a wallet; the victim does not harm by taking it back.
13. There are counterexamples. If I welcome or encourage someone to enter my property but they do not know this and they enter – they believe – against my wishes, and then injure themselves, it seems that I am not *causally* responsible for their injury (or not because of my invitation).
14. Of course, they may both be.
15. In Christian tradition, the Parable of the Good Samaritan is told by Jesus in response to the question of what one must do to be granted eternal life. (It is recorded in chapter 10 of the Gospel of Luke.) Jesus tells of a traveller attacked and left for dead on the road from Jerusalem to Jericho. A priest and a Levite pass him by, but a Samaritan bandages his wounds and puts him on a donkey. The Samaritan takes the man to an inn, cares for him, and pays for his stay. To be granted eternal life, Jesus says, one must do likewise, and love one's neighbour. The significance of the rescuer's identity is that Jesus' Jewish audience had a hostile relationship with the Samaritans – but thought highly of priests and Levites.
16. Issues of non-identity also enter, as our actions in aiding living animals will lead to the creation of animals who would not otherwise exist. The duties that we possess might translate into duties to the descendants of those we had initial obligations to – whoever these descendants turn out to be.
17. There are two complicating factors. First, many people feel some disgust towards rats, meaning that this may do some work in the intuitive plausibility of excluding rats. Second, some people *do* welcome rats, even if it may be much less common than welcoming (say) garden birds. It is hard to think of urban animals who would *not* be welcomed by at least some people. Small fish or (sentient) invertebrates living in urban waterways may provide examples of animals neither welcomed nor considered aversive.
18. Some urban animals, though, have 'internal dependencies' (Palmer, 2010, p. 92), meaning they have little prospect of life away from humans. The existence of these internal dependencies is surely itself normatively significant.
19. This answer is too quick – many would suffer and die in the transition.
20. A stark example is the introduction of corncrake-friendly mowing practices. Plowing from the inside of fields moving outwards rather than the reverse, for example, is more corncrake friendly – it gives chicks chance to escape, especially when combined with green spaces around the edge of fields. This is something that some Scottish farmers, working with the Royal Society for the Protection of Birds, have adopted, leading to an increase of corncrake numbers. See Barrett (2017).

Acknowledgments

This paper was originally published as an open-access paper in the *Journal of Agricultural and Environmental Ethics*. The paper has been lightly updated. I offer thanks to the editors of *Ethics, Policy, & Environment* for being willing to include it in this special issue, for which it was originally written. I also offer thanks to Kyle Johannsen, Alasdair Cochrane, and Gary O'Brien, who offered very useful comments on earlier drafts of this paper, and to two anonymous reviewers for the *Journal of Agricultural and Environmental Ethics*, who helped me refine my argument. I also offer thanks to the British Academy, who funded my research at the time this article was originally written.

Disclosure Statement

No potential conflict of interest was reported by the author.

Funding

The work was supported by the British Academy [grant number PF19\100101].

ORCID

Josh Milburn http://orcid.org/0000-0003-0638-8555

References

ACE. (2022, November). *Recommended charities*. Animal Charity Evaluators. https://animalcharityevaluators.org/donation-advice/recommended-charities/

Barrett, J. (2017). *Five species we're helping with farmers and crofters*. Royal Society for the Protection of Birds. https://community.rspb.org.uk/ourwork/b/scotland/posts/five-species-we-39-re-helping-with-farmers-and-crofters

Blattner, C. (2020). Animal labor, ecosystem services. *Animal & Natural Resource Law Review, XVI*, 1–40.

Boisvert, R. (2014). Hospitality and food. In P. B. Thompson & D. M. Kaplan (Eds.), *Encyclopedia of food and agricultural ethics* (pp. 1182–1190). Springer.

Brown, A. (2017). A theory of legitimate expectations. *The Journal of Political Philosophy, 25*(4), 435–460. https://doi.org/10.1111/jopp.12135

Browning, H., & Veit, W. (2023). Positive wild animal welfare. *Biology & Philosophy, 38*(14), 1–19. https://doi.org/10.1007/s10539-023-09901-5

Cochrane, A. (2018). *Sentientist politics: A theory of global inter-species justice*. Oxford University Press.

Cooke, S. (2019). Betraying animals. *The Journal of Ethics, 23*(2), 183–200. https://doi.org/10.1007/s10892-019-09289-z

Delon, N., & Purves, D. (2018). Wild animal suffering is intractable. *Journal of Agricultural & Environmental Ethics, 31*(2), 239–260. https://doi.org/10.1007/s10806-018-9722-y

Donaldson, S., & Kymlicka, W. (2011). *Zoopolis: A political theory of animal rights*. Oxford University Press.

Ebert, R., & Machan, T. R. (2012). Innocent threats and the moral problem of carnivorous animals. *Journal of Applied Philosophy*, *29*(2), 146–159. https://doi.org/10.1111/j.1468-5930.2012.00561.x

Faria, C. (2023). *Animal ethics in the wild: Wild animal suffering and intervention in nature*. Cambridge University Press.

Fischer, B. (2022). Two challenges to Johannsen on habitat destruction. *Philosophia*, *50*(3), 865–873. https://doi.org/10.1007/s11406-021-00416-5

Francis, L. P., & Norman, R. (1978). Some animals are more equal than others. *Philosophy*, *53*(206), 507–527. https://doi.org/10.1017/S0031819100026358

Horta, O. (2017). Animal suffering nature: The case for intervention. *Environmental Ethics*, *39*(3), 261–279. https://doi.org/10.5840/enviroethics201739320

Jalagania, B. (2021). Wild animals and duties of assistance. *Journal of Agricultural & Environmental Ethics*, *34*(2), 1–15. https://doi.org/10.1007/s10806-021-09854-3

Johannsen, K. (2021). *Wild animal ethics: The moral and political problem of wild animal suffering*. Routledge.

Lamey, A. (2019). *Duty and the beast: Should we eat meat in the name of animal rights?* Cambridge University Press.

Machan, T. (2004). *Putting humans first: Why we are nature's favorite*. Rowman & Littlefield.

Milburn, J. (2018). Nozick's libertarian critique of Regan. *Between the Species*, *21*(1), 68–93.

Milburn, J. (2022). *Just fodder: The ethics of feeding animals*. McGill-Queen's University Press.

Ng, Y. (1995). Towards welfare biology: Evolutionary economics of animal consciousness and suffering. *Biology & Philosophy*, *10*(3), 255–285. https://doi.org/10.1007/BF00852469

O'Brien, G. D. (2021). Beneficence, non-identity, and responsibility: How identity-affecting interventions in nature can generate secondary moral duties. *Philosophia*, *50*(3), 887–898. https://doi.org/10.1007/s11406-021-00419-2

Palmer, C. (2003). Placing animals in urban environmental ethics. *Journal of Social Philosophy*, *34*(1), 64–78. https://doi.org/10.1111/1467-9833.00165

Palmer, C. (2010). *Animal ethics in context*. Columbia University Press.

Palmer, C. (2021). Should we provide the bear necessities? Climate change, polar bears and the ethics of supplemental feeding. In B. Bovenkerk & J. Keulartz (Eds.), *Animals in our midst: The challenges of co-existing with animals in the Anthropocene* (pp. 377–398). Springer.

Platt, P. (2007). Vertical farming: An interview with Dickson Despommier. *Gastronomica*, *7*(3), 80–87. https://doi.org/10.1525/gfc.2007.7.3.80

Regan, T. (2004). *The case for animal rights* (updated ed.). University of California Press.

Sachs, B. (2019). Teleological contractarianism. *Journal of Social Philosophy*, *50*(1), 91–112. https://doi.org/10.1111/josp.12266

Sagoff, M. (1984). Animal liberation and environmental ethics: Bad marriage, quick divorce. *Osgoode Hall Law Journal*, *22*(5), 297–307.

Scheffler, S. (1997). Relationships and responsibilities. *Philosophy & Public Affairs*, *26*(3), 189–209. https://doi.org/10.1111/j.1088-4963.1997.tb00053.x

Sebo, J. (2022). *Saving animals, saving ourselves: Why animals matter for pandemics, climate change, and other catastrophes*. Oxford University Press.

Smith, K. K. (2012). *Governing animals: Animal welfare and the liberal state*. Oxford University Press.

Telfer, E. (1996). *Food for thought: Philosophy and food*. Routledge.

Valentini, L. (2013). Canine justice: An associative account. *Political Studies*, *62*(1), 37–52. https://doi.org/10.1111/j.1467-9248.2012.01006.x

Zamir, T. (2007). *Ethics and the beast: A speciesist argument for animal liberation*. Temple University Press.

OPEN ACCESS

The Rebugnant Conclusion: Utilitarianism, Insects, Microbes, and AI Systems

Jeff Sebo

ABSTRACT

This paper considers questions that small animals and AI systems raise for utilitarianism. Specifically, if these beings have more welfare than humans and other large animals, then utilitarianism implies that we should prioritize them, all else equal. This could lead to a 'rebugnant conclusion', according to which we should, say, create large populations of small animals rather than small populations of large animals. It could also lead to a 'Pascal's bugging', according to which we should, say, prioritize large populations of small animals even if they have a low chance of being sentient. I suggest that utilitarians should accept these implications in theory but might be able to avoid some of them in practice.

1. Introduction

We are currently in the midst of rapid moral circle expansion. Animal advocates have made significant progress over the past fifty years by promoting the idea that we have moral duties to domesticated animals. We are now in the early stages of promoting the idea that we have moral duties to wild animals. Some of us accept that we have such duties because we think that we should help others when we can. Others of us accept that we have such duties because we think that we are harming many of these animals, and that we should reduce and repair these harms when we can. Regardless, the idea that we have duties to many nonhuman animals is fast gaining acceptance (Johannsen, 2021; Palmer, 2010; Sebo, 2022).[1]

This moral circle expansion raises many difficult questions about our moral priorities. For instance, humans are currently harming and killing tens of billions of domesticated animals per year and hundreds of billions of wild animals per year (Sanders, 2020). At least in terms of scale and neglectedness, then, our duties to current and near future nonhuman animals would seem to take priority over our duties to current and near future humans, all else equal. Granted, we might think that we should prioritize humans for other reasons, including reasons involving tractability and indirect effects, as we will see. Still, we are slowly coming to terms with the idea that nonhumans matter much more than we previously thought.

But as substantial as this moral circle expansion has been, it is not nearly complete. For instance, our discussion of duties to captive and domesticated animals tends to focus on animals such as cats, dogs, cows, pigs, and chickens. And our discussion of duties to free and wild animals tends to focus on animals such as chimpanzees, elephants, koalas, dolphins, and polar bears. While there is no single category that includes all these animals, in general we seem to focus more on large animals than on small animals, on vertebrates than on invertebrates, and on land animals than on aquatic animals. The result is a moral community that is many times larger than it was before, but still many times smaller than, I believe, it should be.

When we take seriously the possibility of a moral community that includes all sentient beings – large and small, vertebrate and invertebrate, terrestrial and aquatic – we realize that this next expansion might, if anything, be even more transformative than the last one. The world is full of conflicting interests and needs, and it is also full of very different kinds of populations. For instance, if we have to choose between improving the lives of a small number of large animals and improving the lives of a large number of small animals, then which should we choose and why? This kind of case requires us to think not only about what we owe each kind of animal but also about how to compare these duties when they conflict.

Suppose that we determine that large animals like humans have more welfare on average but that small animals like insects have more welfare in total. What follows for ethics and politics? Which populations should we prioritize within each generation, all else equal? And which populations should we prioritize across generations, all else equal? Suppose further that we determine that many beings, including microscopic organisms and current and near future AI systems, are at least *possibly* sentient, and that the size of these populations relative to insects rivals the size of insect populations relative to humans. How, if at all, should that possibility affect our moral priorities within and across generations?

My aim in this paper is to survey these questions from a utilitarian perspective, building on work from Horta (2010), Ng (1995), Tomasik (2015), and others. I will show that utilitarianism implies that insects can take priority over humans and that microbes or AI systems can likewise take priority over insects. Granted, we might still have reason to prioritize humans at present and in the near future, since our duties regarding the distant future outweigh our duties regarding the present and near future, and since improving human lives at present and in the near future is key to improving the distant future. But in this case, what saves utilitarians from one surprising conclusion might be another surprising conclusion.

To be clear, while I focus on utilitarianism here for the sake of simplicity, I think that other moral theories face versions of these questions as well. Any theory that involves a duty of beneficence or a duty of non-maleficence will have to deal with questions about, say, how to set priorities between small populations of large animals and large populations of small animals, since there might be many cases where we have the power to help or avoid harming *either* the former populations *or* the latter populations but not both at the same time. So while my discussion here might focus on how one moral theory might navigate this strange future, we should keep in mind that this strange future awaits us all.

2. Background

Any discussion about how to set priorities must begin with a certain set of normative and empirical assumptions. So I begin in this section by clearly stating the assumptions that will govern my discussion here. Normatively, I will assume a classical utilitarian theory of morality, according to which we morally ought to maximize positive welfare in the world.[2] And empirically, I will assume that many beings might have the capacity for welfare, that beings with larger brains and longer lifespans tend to have a higher capacity for welfare than beings with smaller brains and shorter life spans on average, and that many other factors will be relevant to our priorities too, such as tractability and indirect effects.

Consider first my normative assumptions.[3] Utilitarianism, as I will be interpreting it here, involves a *hedonist* theory of the good. According to hedonism, the only intrinsically and finally good state is pleasure, that is, positively valenced conscious experience. And the only intrinsically and finally bad state is pain, that is, negatively valenced conscious experience. Everything else is only, at most, extrinsically or instrumentally valuable, depending on its relationship with pleasure or pain. For example, to the degree that knowledge promotes more pleasure than pain, knowledge is good according to hedonism. But to the degree that knowledge has the opposite effect, knowledge is bad according to hedonism.[4]

Utilitarianism, as I will be interpreting it here, also involves a *totalist* theory of the good, according to which the world is better when it contains more pleasure and less pain in total. For example, if one world contains more pleasure in total and another contains more pleasure for the worst-off, utilitarianism implies that the former world is better. Of course, we might still have reason to prioritize the worst-off in many cases on this view, since, for instance, an additional $1,000 will tend to make a bigger difference for low-income person than for a high-income person and, so, distributing this money to the low-income person will tend to do more good, all else equal. Either way, what does the most good is best.

Finally, utilitarianism, as I will be interpreting it here, involves a *maximizing act consequentialist* theory of the right. According to act consequentialism, the rightness or wrongness of an action depends on the consequences of that action. And according to maximizing act consequentialism, an action is right if and only if it maximizes the good (or, in the scalar formulation, an action is right to the degree that it maximizes the good). And when we combine maximizing act consequentialism with hedonism and totalism, we get classical utilitarianism, as I will be interpreting it here: An action is right if and only if, or to the degree that, it produces the most pleasure and least pain possible, in total.

At least in theory, utilitarianism is highly demanding. It implies that we might sometimes be required to sacrifice our own well-being for the sake of the greater good or sacrifice the few for the sake of the many. Granted, we might think that there are limits to how demanding utilitarianism can be in practice, since, for instance, we need to take care of ourselves to be able to take care of others, and we need to treat others as having rights to maintain the kinds of norms and structures that allow us to maximize pleasure in the long run. But these considerations might not always be decisive, and even when they are, utilitarianism might still be more demanding than commonsense morality in many cases in practice.[5]

As this last point suggests, I am not assuming utilitarianism here because I reject other moral theories. Far from it. As I argue elsewhere (Sebo, 2022), I think that most people in most situations should accept a pluralistic, partly consequentialist and partly non-consequentialist moral theory in practice. For instance, utilitarians should accept such a theory since we sometimes need to respect rights to promote welfare, and rights theorists should accept such a theory because we sometimes need to promote welfare to respect rights. We should also accept such a theory on strategic grounds, since it can serve as the basis for collaboration between consequentialists and non-consequentialists in advocacy and policy.

Instead, I am assuming utilitarianism here simply because the issues that I discuss in this paper are complex, and it would be difficult to assess them from multiple moral theories at once. So my strategy will be to (start to) assess these issues from a utilitarian perspective for the sake of simplicity and specificity. My hope is that we can then assess these issues from other perspectives as well, and that we can then put these analyses together to arrive at a preliminary view about how to set priorities in a multi-species community that includes small populations of large animals and large populations of small animals (and other possibly sentient beings). To that extent, my analysis here will be highly preliminary and incomplete.

Now consider my empirical assumptions. My first empirical assumption is that many beings at least *might* be sentient (that is, might have the capacity for welfare), given the evidence available. Granted, we might think that some beings are more likely to be sentient than others. For instance, we might think that mammals are more likely to be sentient than insects, and that insects are more likely to be sentient than microbes or current AI systems. But we are not, at present, able to rule out the possibility that any of these beings are sentient. Even if we assign an *astronomically low* probability to the idea that, say, microbes are sentient, we should not assign a probability of *zero* to that idea at present.

My reason for making this assumption is that the problem of other minds limits how much we can know about other minds at present and for the foreseeable future. Since the only mind that I can directly access is my own, I am not able to directly confirm or disconfirm what, if anything, it might be like to be anyone or anything other than myself. And while some theories of consciousness, such as higher order thought theories, imply that only some of the beings listed above are conscious, other theories, such as panpsychist theories, imply that they all are. Unless and until we make substantial progress on this problem, we will need to keep an open mind about the scope of sentience in the world (Carruthers, 2004; Sebo, 2018).

My second empirical assumption, which I gestured at a moment ago, is that beings with larger brains and longer life spans will tend to have a higher capacity for welfare than beings with smaller brains and shorter life spans. For instance, we might think that an individual human can experience more pleasure and pain than an individual insect, and that an individual insect can experience more pleasure and pain than an individual microscopic organism (assuming, of course, that these beings can experience any pleasure and pain at all). Granted, we might think that some of these beings experience *astronomically little* pleasure and pain, if any at all. But if they are sentient, then they might experience at least *some*.

My reason for making this empirical assumption is that it seems plausible that an individual with more complex neural systems related to pleasure and pain would be capable of a higher amount of pleasure and pain at a time, all else equal. It also seems plausible that an individual with a longer life span would be capable of a higher amount of pleasure and pain over time, all else equal. Granted, there are many complications here, and it is unlikely that the capacity for welfare at a time will be a simple function of neuron counts, or that capacity for welfare over time will be a simple function of life spans (Schukraft, 2020). But this general assumption is all I need to motivate the questions that I will ask in this paper.

My third and final empirical assumption is that other factors, such as tractability and indirect effects, are relevant to which beings we should prioritize according to utilitarianism. First, we need to consider the tractability of the problems that we face. For instance, even if our actions impact many insects, we might not have a duty to help these insects in practice if we have *no knowledge at all* about how to help them or *no power at all* to help them in particular contexts. I will suggest that we should, in fact, limit how much we prioritize insects and other such beings at present for these reasons. But I will also suggest that we should still prioritize these beings much more than we do, and much more in the near future than at present.

Second, we need to consider the indirect effects of our solutions to these problems. In particular, even if our impacts on near future insects matter more than our impacts on near future humans *all else equal*, our impacts on near future humans might matter more than our impacts on near future insects *all things considered*, since what happens to humanity will likely determine how many sentient beings can exist and how good their lives can be in the long run. I will once again suggest that we should, in fact, limit how much we prioritize insects and other such beings for these reasons. But I will also, once again, suggest that we should still prioritize them much more than we do, and much more in the near future than at present.

If we put these normative and empirical assumptions together, then we find that utilitarianism might be an even more radical departure from commonly accepted moral norms than we expected. At least in principle, it implies that we can have a duty to prioritize insects over humans and microbes and AI systems over insects. And while utilitarians might be able to avoid some of these implications in practice given the numbers involved or considerations involving tractability and indirect effects, we will likely not be able to avoid *all* of them in these ways. No matter what, a commitment to doing the most good possible will likely require us to prioritize beings who are very unlike us, either in the short term or in the long run.

3. The Rebugnant Conclusion

Utilitarianism might have surprising implications about our duties regarding future insect populations. Plausibly, maximizing utility in the future partly requires creating a world in which the global population can flourish in relative harmony. But it also partly requires considering which beings to add to this global population in the first place. For instance, should we aim to bring about a world with a higher ratio of large to small animals, a higher ratio of small to large animals, or a balance between the two? While asking such questions might seem like playing God, the reality is that humans are already shaping future populations whether we like it or not. We need to be thoughtful about how we wield this power.

This question is related to a problem in population ethics that Derek Parfit calls *the repugnant conclusion* (Parfit, 1984, pp. 381–90). Here is the problem. Suppose that we have to choose between two future populations: Population A contains one million people with one million units of pleasure each, and population B contains two million people with 999,999 units of pleasure each. Which one should we choose? Many people have the intuition that we should choose B, since there are twice as many people and everyone is nearly as happy on average, and there is nearly twice as much pleasure overall. This intuition is friendly to utilitarianism, which implies that we should maximize total utility, not average utility.

So far so good. But now suppose that we take this reasoning farther. Instead of choosing two million people with 999,999 units of pleasure each, we choose four million with 999,998 each. Then we choose eight million with 999,997 each. Eventually we choose a population where everyone has only one unit of pleasure, but where the population is so large that they still have nearly twice as much pleasure as the previous one overall. Many people have the intuition that we should *not* choose such a population; in fact, Parfit goes so far as to call the idea that we should choose this population *the repugnant conclusion*. This intuition – which, to be fair, not everyone shares (see, for instance, Ng, 1989) – is less friendly to utilitarianism.

While many people default to thinking about the repugnant conclusion in same-species cases, we can think about it in multispecies cases as well. In a same-species case, we might compare small human populations with high average welfare with large human populations with low (but still positive) average welfare. In a multispecies case, we might compare small human populations with high average welfare with large, say, ant populations with low (but still positive) average welfare. Either way, the outcome is the same in both cases: One population contains much more pleasure on average, the other population contains much more pleasure in total, and we need to determine which state of affairs is better all else equal.

Granted, there are many relevant differences between same-species and multispecies cases, particularly in the real world. For instance, we might wonder whether we have to choose between humans and ants at all and whether ants really do experience more pleasure than humans in total. But while these details might allow us to avoid this problem in some cases in practice, they do not allow us to avoid the problem in principle, or even, necessarily, in all cases in practice. *If* we have to choose between humans and ants, and *if* the ants would experience more pleasure than humans in total, *then* we should bring about the ant population all else equal, according to this view. We can call this implication *the rebugnant conclusion*.[6]

In general, when a seemingly plausible argument leads to a seemingly implausible conclusion, we have two options. First, we can accept the conclusion. For instance, in this case we can accept that we might sometimes be required to produce large populations of small animals rather than small populations of large animals, provided that doing so will, in fact, maximize utility. Granted, we might experience this implication as implausible. But all moral theories have *some* implications that we might experience as implausible. And given our self-interest, speciesism, status quo bias, scope insensitivity, and so on, we should expect to experience some implications as implausible even if they are in fact correct.

Our second option is to reject features of utilitarianism that lead to this conclusion. For example, if we reject hedonism, then we might be able to deny that human and ant well-being are comparable (see, for instance, Korsgaard, 2018). If we reject totalism, then we might be able to deny that the world with the most pleasure in total is best.[7] And if we reject maximization, then we might be able to deny that we are morally required to produce the best world.[8] Philosophers commonly make such moves to reject the repugnant conclusion, and they can do the same here. My own view is that accepting this conclusion is better than rejecting some or all of these features of utilitarianism, but others might disagree.[9]

But even if we accept this conclusion in principle, we might be able to reject or constrain it in many cases in practice. First, we might worry about tractability. It seems relatively tractable to create more humans and other large animals *and* ensure that they have net positive experiences in the future. In contrast, while it seems similarly tractable to create more insects and other small animals in the future, it seems less tractable to ensure that they have net positive experiences, particularly insofar as they continue to reproduce through r-selection, that is, the evolutionary strategy that involves having a relatively large number of babies with a relatively high infant mortality rate (Groff & Ng, 2019; Horta, 2010; Tomasik, 2015).

Second, we might worry about the indirect effects of prioritizing large populations of small animals for the far future. That is, even if we feel confident that increasing the ratio of small to large animals would maximize utility in the short term, we might not be confident that it will do so in the long run. The reason is that if we want to maximize utility in the far future, then we might need to ensure that humanity can survive long enough to ensure that many beings will exist in the far future and have net positive welfare. This consideration might save us from a surprising neartermist implication, but it would do so only by replacing it with a surprising longtermist implication, as we will discuss in the next two sections.

Finally, we might worry about empirical facts regarding insects and other small animals. For us to face the rebugnant conclusion in the real world, it would need to be the case that we face a choice between bringing about a small population of large animals and bringing about a large population of small animals, that the population of small animals would have less welfare on average and more welfare in total than the population of large animals, and that this population of small animals would have net positive welfare rather than net negative welfare. If we reject any of these assumptions, then we might be able to avoid the rebugnant conclusion, at least in its pure form, in the real world in the near future.

My own view is that the first two points, about tractability and indirect effects, are plausible but not decisive. It seems clear that we can more effectively improve the lives of small populations of large animals than improve the lives of large populations of small animals at present, due to our epistemic and practical limitations. But this situation can change. The more we improve our knowledge about small animals and our power to help them, the more we might be able to improve their lives effectively at scale as well. And if and when that happens, utilitarians will have to take more seriously the possibility that we should prioritize expanding and assisting these populations on grounds of scale and tractability.

I also think that the third point, about empirical facts regarding these animals, is plausible in this context. At least for now, all realistic options will involve bringing about mixed populations of large and small animals with good and bad lives. But as we will see in §5, we might face a choice between increasing or decreasing the ratio of large to small animals in the world. And whether or not small animals have net positive welfare, we should still consider them when deciding what to do. Specifically, if they have net positive welfare, then we should treat any policy that creates more or fewer of them as good or bad, respectively, to that degree. And if they have net negative welfare, then we should do the reverse.

With all that said, my guess is that at the end of the day, humans will not be required to prioritize bringing about insects and other small animals in the near future for reasons involving tractability, indirect effects, and uncertainty about their welfare, according to utilitarianism. But I should note two caveats. First, even if we might not be required to prioritize bringing about insects and other small animals in the near future, we might still be required to consider them much more than we do. For instance, to the degree that we expect practices such as agriculture, deforestation, and development to create more or fewer small animals with good or bad lives, we should assign weight to this consideration when deciding what to do.

Second, and additionally, even if humans are morally permitted to prioritize bringing about ourselves and other large animals in the near future, we might not be morally permitted to do so in the far future, according to utilitarianism. As I have mentioned and will discuss more in the next section, insofar as we take a longtermist perspective, it might be that we should prioritize bringing about very different kinds of beings in the far future. And either way, it might be that we should help insects and other small animals much more than we are at present as a means to this end, since helping these animals now is part of what will allow humanity to expand our moral circle and treat a wider range of nonhuman beings better in the future.

4. Pascal's Bugging

Utilitarianism might also have surprising implications about our duties regarding future microbe populations. In my view, the real test is not beings who have, say, a >10% chance of being sentient, but rather beings who have, say, a >.000001% chance of being sentient. Specifically, the real test is the prospect of an *astronomically high* number of beings who have an *astronomically low* but non-zero chance of being sentient. In this kind of case, as long as this population is large enough, utilitarianism (coupled with standard decision theory) can imply that we should favor it all else equal, even if its members all have a *very low* chance of being sentient and a *very low* level of welfare if any at all.

This problem is an instance of what Eliezer Yudkowsky calls *Pascal's mugging* (Yudkowsky, 2007 see also Bostrom, 2009). Suppose that a mugger walks up to you and proposes a deal: If you give them $10 today, then they promise to invest your money, make a profit, and give you $20 tomorrow. Plausibly, you have reason to decline this offer, since you have reason to doubt each part of this proposal: that the mugger would invest your money, that they would make a profit even if they did invest it, and that they would pay you back even if they did make a profit. And as long as your credence that the mugger will keep this promise is lower than .5, giving them $10 is not worthwhile in expectation, all else equal.

So far so good. But now suppose the mugger keeps negotiating. They promise to return with increasingly high amounts of money, and they explain to you how they would convert your money into these higher amounts. With each new promise, you might think that the mugger is less likely to keep this promise but that it would be better if they did. In this case, as long as the increase in benefit sufficiently outpaces the decrease in probability (and the probability remains non-zero), at some point you might become rationally required to take the deal. This implication presents a challenge to standard decision theory. Can we really be rationally required to bet on an astronomically low chance of an astronomically high benefit?

Similar problems can arise for utilitarianism. Suppose we have an astronomically low chance of producing an astronomically high number of distant future happy people. In this case, as long as this population is large enough, we might be morally required to prioritize producing this population according to utilitarianism, all else equal. That is, utilitarianism might require us to pursue a course of action that has an astronomically low chance of creating and benefiting an astronomically high number of far future people, rather than courses of action that have a much higher chance of creating and/or benefiting much lower (but still high) numbers of current or near future people (for relevant discussion, see Greaves & MacAskill, 2021).

Now consider a similar case. Suppose that there is an astronomically low chance that microbes are sentient, and that each microbe can experience an astronomically low amount of welfare if they are. As before, as long as this population is large enough, we might be morally required to prioritize this population according to utilitarianism, all else equal. The math in this case is the same as the math in the previous case. The only difference is that, in the previous case, the subjects in question are definitely sentient but very unlikely to exist, whereas in this case, the subjects in question definitely exist but are very unlikely to be sentient (and have a very low capacity for welfare, if any at all). We can call this implication *Pascal's bugging*.

We can also imagine cases that involve both features at once. For instance, suppose that there is an astronomically low chance that we can bring about an astronomically high number of, say, *distant future microbes*. Suppose further that each microbe has an astronomically low chance of being sentient and has an astronomically low amount of welfare if any at all. Finally, suppose that we can still maximize expected utility by pursuing this option, given the sheer size of this population. In this case, a utilitarian might be required to forego creating and benefiting beings who are *very likely* sentient and/or who will *very likely* exist, all for the sake of beings who are merely *possibly* sentient and who will merely *possibly* exist.

As these examples illustrate, Pascal's bugging raises the stakes of the rebugnant conclusion in two related ways. First, it raises the stakes for intragenerational priority-setting questions because it implies that our decisions about how to set priorities within any particular generation depend not only on the implications for, say, insects, but also on the implications for even larger populations of even smaller beings. For example, if we discover that the world contains so many microbes that the 'micro-population' has more expected welfare than the 'macro-population', then we would have to accept that the rightness or wrongness of our actions depends primarily on the expected impacts on the micro-population, all else equal.

Second, Pascal's bugging raises the stakes for intergenerational priority-setting because it implies that we might need to prioritize not only creating large populations of beings with a low amount of welfare on average, but also *very* large populations of beings with a *very* low chance of being sentient at all and a *very* low amount of welfare on average, if any at all. For example, if our choice is between creating a given number of future mammals, a quintillion times as many future insects, and a septillion times as many future microbes, then, depending on the probabilities and utilities we assign in each case, we might be morally required select not only the insects instead of the mammals but also the microbes instead of the insects.

As before, when faced with this kind of conclusion, we have several options. First, we can accept the conclusion. If current or future micro-populations did, in fact, contain more expected welfare than current or future macro-populations, then we would, in fact, be required to prioritize these mico-populations all else equal. Granted, this implication might seem highly implausible. But as noted above, every moral theory has some implications that we might find implausible, and the point of moral theory is to improve our moral thinking, not to merely confirm what we already think. And in this case, we can expect that human bias and ignorance would prime us to find Pascal's bugging highly implausible even if it was correct.

Alternatively, we can reject features of utilitarianism or standard decision theory that lead to this conclusion. For example, we can once again consider rejecting either the hedonism, totalism, or maximization of utilitarianism. Alternatively, we can consider rejecting the *fanaticism* of standard decision theory, according to which we should consider all possibilities when deciding what to do, including possibilities involving astronomically low probabilities of astronomically high impacts. Rejecting this idea might allow us to discount or bracket astronomically low probabilities of astronomically high impacts when estimating how to do the most good possible (for discussion, see Wilkinson, 2022).

Of course, as with the rebugnant conclusion, even if we accept Pascal's bugging in principle, we might still be able to reject or constrain it in many cases in practice. For example, if we think that we have no knowledge or power regarding microbe welfare, then we might think that we can bracket our impacts on microbes in practice *whether or not* they have more expected value than other animals. But a lot depends on whether we have *zero* knowledge and power or, rather, *very little* knowledge and power. If we have *zero* knowledge and power, then this reply might work. But if we instead have *very little* knowledge and power, then while this reply might add an extra line to our math, it might or might not alter the outcome.

Similarly, if we think that the probability that microbes are sentient is *zero*, then we might be able to avoid this problem with respect to microbe populations. Or if we think that the probabilities and utilities are so low that other populations still have more expected welfare in total, despite the size of microbe populations, then we might be able to avoid the problem with for this reason as well. Given the problem of other minds, I think that the 'low expected welfare' response is more plausible than the 'no expected welfare' response. It would be rash to think that there is *no chance at all* that microbes can consciously experience *any positive or negative states at all*, no matter how minimal. But both responses can be explored.

While there is too much uncertainty for us to say much with confidence at this stage, we have at least some reason to think that microbe populations could carry a lot of weight in expectation. For instance, according to one recent estimate, 'there are between 10^23 and 10^24 neurons on earth', most of which are 'distributed roughly evenly among small land arthropods, fish, and nematodes, or possibly dominated by nematodes'.[10] Nematodes are microscopic organisms with about 300 neurons each. In my view, they are good candidates for the kind of being that we are considering: very low but non-zero chance of being sentient, very low capacity for welfare if any at all, but numerous enough that it could all add up.

As before, we might or might not think that we should *prioritize* nematodes in practice any time soon, if any time at all, according to utilitarianism. But even if not, we might still think that we should at least *consider* them (along with many other possibly sentient beings, as we will see in a moment) much more than we are. In this case we can make all the same points as before: A lot depends on the tractability and indirect effects of promoting nematode welfare, as well as on the exact probabilities and utilities involved. But if and when we do determine that nematodes have net positive or negative expected welfare, we might once again have to treat any policy that creates more or fewer of them as good or bad to that degree.

5. Bugs in the System

Throughout this discussion I have emphasized that utilitarianism might have these implications not only in principle but also, at least to a degree, in some cases in practice. I now want to consider two possible scenarios where that might be true. In the first scenario, utilitarianism could imply that the rightness or wrongness of our actions depends primarily on the expected impacts for large populations of small animals. In the second scenario, it could imply that the rightness or wrongness of our actions depends primarily on the expected impacts for large populations of AI systems. Both of these scenarios are possible, and they could at least partly match the scenarios that we have been considering.

Consider first the scenario involving small animals. Many humans think that global changes such as climate change are bad because of the expected impacts for humans. Of course, this view makes sense. Climate change will cause melting ice caps, rising sea levels, flooding coastal areas, conflict over land, water, and energy, and an increase in extreme weather events such as hurricanes and tsunamis. It will also be a threat multiplier that amplifies existing threats for humans such as hunger, thirst, illness, and injury. As a result, many humans will suffer and die, including and especially the most vulnerable among us. While many humans will also prosper, we can reasonably expect that the net effects for humans will be negative.[11]

But when we consider nonhumans too, we realize that the impacts of climate change will be much more complex. While many animals will suffer, many other animals will flourish. And while many nonhuman populations will contract, many others will expand. While much remains unclear, one possibility – which, to be clear, is only a possibility – is that climate change will produce a world with a higher ratio of small to large animals.

Warmer climates tend to favor smaller animals, both by allowing them to migrate north and south and by causing some larger animals to shrink. As a result, a world reshaped by climate change could contain more animals and higher welfare in total but smaller animals and lower welfare on average.[12]

If this scenario arises – which, again, is a big 'if' – then a lot will depend on whether these small animals have net positive or negative expected welfare. If they have net positive expected welfare, then climate change could produce a real-life rebugnant conclusion, by producing larger populations of smaller animals, such that they experience less pleasure on average but more in total. Otherwise it could have the reverse effect. Either way, if these small animals have more expected welfare than humans and other large animals during this period, then the goodness or badness of climate change during this period will depend primarily on the impacts for these small animals, according to utilitarianism.[13]

Now consider the scenario involving AI systems. If humanity survives climate change and the other global threats that will imperil our species in the near term, then we might face an open future with opportunities for expansion. Specifically, we might have the ability to expand beyond the planet, and we might also have the motivation to do so. After all, as long as we can survive as a species, we likely have at least a billion years before water loss and thermal runaway make this planet uninhabitable (Wolf & Toon, 2014). This gives us plenty of time to develop the knowledge and power necessary to expand. We then have billions of years to spend potentially extending sentient life throughout the universe (Bostrom, 2003).

The question is how we would go about extending sentient life throughout the universe, if we could. Consider two options. The first is a biological approach: We aim to create as many happy biological minds as possible, by building space stations or terraformed planets on which (post-)human and nonhuman animals can live. The second is an artificial approach: We aim to create as many happy artificial minds as possible, by building large, solar-powered networks of artificial minds who can experience pleasure. While we can plausibly achieve a very large population either way, we can plausibly achieve a much larger population and much better outcomes for its members with the artificial approach (John, 2021).

I imagine that utilitarian agents might pursue a mixed approach at first, if for no other reason than to further reduce existential risk and oversee whatever further expansions might be optimal. But I can also imagine scenarios where utilitarian agents pursue an artificial-centric approach beyond that. This approach might or might not fit the structure of the rebugnant conclusion in at least some respects. For instance, depending on the details, it might or might not be that future artificial minds have a lower chance of being sentient and a lower capacity for welfare than future biological minds. But even if they do, the astronomically large size of a future artificial population could be more than enough to make up for that.

Somewhat ironically, it might be that utilitarians can avoid the first 'repugnant' conclusion in part by embracing the second one. That is, even *if* we were to determine that climate change is net positive in the short term (because it creates more small animals and higher total welfare), we might still determine that climate change is net negative in the long run (because it increases existential risks for humanity, thereby reducing the probability that we can survive long enough to spread positive welfare beyond this planet). In this scenario, utilitarianism would vindicate the commonsense view that climate change is bad, but it might do so only instrumentally, in service of strange and surprising far future goals.

With that said, as I have emphasized throughout this paper, even if utilitarians should accept these 'repugnant' conclusions in theory, we might be able to reject some of them in practice. As noted above, utilitarians should accept a pluralistic moral framework that considers welfare, rights, virtues, relationships, and more in practice. Moreover, we need to consider many factors in our impact assessments in practice, such as: how likely particular beings are to be sentient, how much welfare these beings are likely to have, to what extent particular actions will increase or decrease their welfare, and to what extent our successors can overcome epistemic, practical, and motivational obstacles that stand in the way of impartiality.

Once we consider all these factors, we might decide that utilitarians are unlikely to face pure versions of the rebugnant conclusion, Pascal's bugging, or other such problems in practice. For instance, we might decide that promoting welfare while respecting rights, cultivating virtuous characters, and cultivating caring relationships requires not making high-stakes decisions through classical utilitarian reasoning alone. And even insofar as we apply classical utilitarian reasoning in these situations, we might decide that climate change, the artificial path, and other such outcomes are not, in fact, better in total but worse on average, in expectation. In these scenarios, we might still face a strange moral future, but the details might seem more plausible.

But we should keep in mind that, insofar as we do avoid a 'repugnant' conclusion, we might face another surprising conclusion instead. In particular, when we ask what kind of future population is best all things considered, the answer will likely *not* be an expanded population of beings like us, but will likely instead be *either* (a) a much larger population of much smaller beings *or* (b) a much smaller population of much larger beings. And whereas answer (a) might remind us of a 'repugnant conclusion' case, answer (b) might remind us of a 'utility monster' case, that is, a case where we can maximize expected utility by prioritizing a relatively small number of beings with a relatively large amount of happiness on average (Nozick, 1974).

We should also keep in mind that, even if an expanded population of beings like us *is* best all things considered, that fact would still seem like a 'repugnant conclusion' case to a much smaller population of much larger beings, and it would still seem like a 'utility monster' case to a much larger population of much smaller beings. While we might be inclined to see human-like populations as the norm and other kinds of populations as deviations, the reality is that every kind of population is a norm from some possible perspectives and a deviation from others. And if we want to think about population ethics in an impartial manner, then we should avoid anchoring to our own contingent perspectives about such matters.

Granted, even if we keep all these considerations in mind, we might still have a hard time shaking the intuition that we should prioritize human-like populations in both the short term and the long run. No matter how much progress we make as a species, we might continue to find it strange that our reward for improving human lives and social, political, and economic systems is that we have an increasingly strong responsibility to live in service of very different kinds of (possibly) sentient being. But these strike me as the kinds of bullets that utilitarians should be willing to bite, especially since we know that human bias and ignorance will likely continue to shape our moral intuitions even as our beliefs, values, and practices improve.

In any case, I think that the immediate implications of this discussion are highly intuitively plausible: We should work at present to build knowledge, power, and political will toward creating a better world for humans and other sentient beings. We might or might not face any pure 'repugnant' or 'utility monster' cases in practice. But if we do, we will need to be epistemically, practically, and motivationally prepared. And even if not, we will still face the weaker – but still radical – conclusion that beings who are very different from us might matter much more than we appreciate. We will need to consider our impacts on these beings very carefully in the future, whether or not we should prioritize them over ourselves.

In closing, it is worth reiterating that many of these problems can arise for non-utilitarian moral theories as well. Whether we think about morality primarily in terms of welfare, rights, virtues, relationships, or other features of life, we will have to ask how to set priorities between large populations of small beings and small populations of large beings, both at present and in the future. And while we might be able to avoid some of the implications discussed here by rejecting features of utilitarianism such as hedonism, totalism, maximization, or fanaticism, we might not be able to avoid all of them that way. Ultimately, the more we accept how large and varied the moral community is, the stranger morality will become.

Notes

1. Thanks to Kyle Johannsen, two anonymous referees, and the organizers and participants of the 8th Oxford Workshop on Global Priorities Research for helpful feedback on previous drafts of this paper.
2. Note that some people use the term 'welfare' in a valenced way, to refer to positive welfare states (in this case we can contrast 'welfare' with 'illfare'), whereas other people use the term 'welfare' in a non-valenced way, to refer to any welfare states (in this case we can contrast 'positive welfare' with 'negative welfare'). This paper will use the term in this latter sense, but nothing substantive will turn on that terminological choice.
3. For more on the features of utilitarianism discussed in this section, see De Lazari-Radek and Singer (1974). For general discussion of the pros and cons of utilitarianism and other moral theories, see Driver (2007).
4. This paper will use 'utility' to refer to goodness in this hedonic sense.
5. For more on this kind of indirect consequentialism, see Sidgwick (1874/2011 and Hare (1981).
6. For similar discussion, see Sebo (2022).
7. For analysis of key assumptions underlying totalism, see Hirose (2014).
8. For discussion of non-maximizing views, see Jamieson and Elliot (2009), Slote (1984), and Sinhababu (2018).
9. For discussion of why the repugnant conclusion is not necessarily repugnant, see Zuber et al. (2021).
10. For informal discussion, see Ray (2017), 'How many neurons are there?' *Eukaryote Writes Blog*: https://eukaryotewritesblog.com/how-many-neurons-are-there/.
11. For more on the nature and ethics of climate change, see Jamieson (2014).
12. For more on the impacts on the distribution of animals, see Bebber et al. (2013). For more on the impacts on the size of animals, see Weeks et al. (2020).
13. We might also think that large populations of small animals are more likely than small populations of large animals to have net negative welfare, since the former populations are more likely to be r-selected and, so, to have higher infant mortality rates. For more, see Johannsen (2021).

Disclosure statement

No potential conflict of interest was reported by the author(s).

References

Bebber, D., Ramotowski, M., & Gurr, S. (2013). Crop pests and pathogens move polewards in a warming world. *Nature Climate Change*, *3*(11), 985–988. https://doi.org/10.1038/nclimate1990

Bostrom, N. (2003). Astronomical waste: The opportunity cost of delayed technological development. *Utilitas*, *15*(3), 308–314. https://doi.org/10.1017/S0953820800004076

Bostrom, N. (2009). Pascal's wager. *Analysis*, *69*(3), 443–445. https://doi.org/10.1093/analys/anp062

Carruthers, P. (2004). *The Nature of the mind: An introduction, Chapter 1*. Routledge.

De Lazari-Radek, K., & Singer, P. (2017). *Utilitarianism: A very short introduction*. Oxford University Press).

Driver, J. (2007). *Ethics the fundamentals*. Blackwell Publishing.

Greaves, H., & MacAskill, W. (2021). *The case for strong longtermism* (GPI Working Paper No. 5). https://globalprioritiesinstitute.org/hilary-greaves-william-macaskill-the-case-for-strong-longtermism-2/

Groff, Z., & Ng, Y.K. (2019). Does suffering dominate enjoyment in the Animal Kingdom? An update to welfare biology. *Biology & Philosophy*, *34*(4), 40. https://doi.org/10.1007/s10539-019-9692-0

Hare, R. M. (1981). *Moral thinking: Its levels, method, and point*. Oxford University Press.

Hirose, I. (2014). *Moral aggregation*. Oxford University Press.

Horta, O. (2010). Debunking the idyllic view of natural processes: Population dynamics and suffering in the wild. *Telos: Revista Iberoamericana de Estudios Utilitaristas*, *17*(1), 73–90.

Jamieson, D. (2014). *Reason in a dark time: Why the struggle against climate change failed - and what it means for our future*. Oxford University Press.

Jamieson, D., & Elliot, R. (2009). Progressive consequentialism. *Philosophical Perspectives*, *23*(1), 241–251. https://doi.org/10.1111/j.1520-8583.2009.00169.x

Johannsen, K. (2021). *Wild animal ethics: The moral and political problem of wild animal suffering*. Routledge.

John, T. (2021). Panspecies longtermism, unpublished manuscript.

Korsgaard, C. M. (2018). *Fellow creatures: Our obligations to the other animals*. Oxford University Press.

Ng, Y.K. (1989). What should we do about future generations? *Economics & Philosophy*, *5*(2), 245–253. https://doi.org/10.1017/S0266267100002406

Ng, Y.K. (1995). Towards welfare biology: Evolutionary economics of animal consciousness and suffering. *Biology & Philosophy*, *10*(3), 255–285. https://doi.org/10.1007/BF00852469

Nozick, R. (1974). *Anarchy, state, and Utopia*. Basic Books.

Palmer, C. (2010). *Animal ethics in context*. Columbia University Press.

Parfit, D. (1984). *Reasons and persons*. Oxford University Press.

Ray, G. (2017). *How many neurons are there?* Eukaryote Writes Blog. https://eukaryotewritesblog.com/how-many-neurons-are-there/

Sanders, B. (2020). *Global animal slaughter statistics and charts: 2020 update*. Faunalytics. https://faunalytics.org/global-animal-slaughter-statistics-and-charts-2020-update/

Schukraft, J. (2020). *Differences in the intensity of valenced experience across species*. Effective Altruism Forum. https://forum.effectivealtruism.org/posts/H7KMqMtqNifGYMDft/differences-in-the-intensity-of-valenced-experience-across

Sebo, J. (2018). The moral problem of other minds. *The Harvard Review of Philosophy*, *25*, 51–70. https://doi.org/10.5840/harvardreview20185913

Sebo, J. (2022). *Saving animals, saving ourselves*. Oxford University Press.

Sidgwick, H. (1874/2011). *The methods of ethics*. Cambridge University Press.

Sinhababu, N. (2018). Scalar consequentialism the right way. *Philosophical Studies*, *175*(12), 3131–3144. https://doi.org/10.1007/s11098-017-0998-y

Slote, M. (1984). Satisficing consequentialism. *Proceedings of the Aristotelian Society, 58*(1), 139–176. https://doi.org/10.1093/aristoteliansupp/58.1.139

Tomasik, B. (2015). The importance of wild animal suffering. *Relations, 3*(2), 133–152. https://doi.org/10.7358/rela-2015-002-toma

Weeks, B. C., Willard, D. E., Zimova, M., Ellis, A. A., Witynski, M. L., Hennen, M., & Winger, B. M. (2020). Shared morphological consequences of global warming in North American migratory birds. *Ecology Letters, 23*(2), 316–325. https://doi.org/10.1111/ele.13434

Wilkinson, H. (2022). In defence of fanaticism. *Ethics, 132(*2): 445–477.

Wolf, E. T., & Toon, O. B. (2014). Delayed onset of runaway and moist greenhouse climates for earth. *Geophysical Research Letters, 41*(1), 167–172. https://doi.org/10.1002/2013GL058376

Yudkowsky, E. (2007). *Pascal's mugging: Tiny probabilities of vast utilities*. Less Wrong. https://www.lesswrong.com/posts/a5JAiTdytou3Jg749/pascal-s-mugging-tiny-probabilities-of-vast-utilities.

Zuber, S., Venkatesh, N., Tännsjö, T., Tarsney, C., Stefánsson, H. O., Steele, K., Spears, D., Sebo, J., Pivato, M., Ord, T., Ng, Y.K., Masny, M., MacAskill, W., Lawson, N., Kuruc, K., Hutchinson, M., Gustafsson, J. E., Greaves, H., Forsberg, L., & Asheim, G. B. (2021). What should we agree on about the repugnant conclusion? *Utilitas, 33*(4), 1–5. https://doi.org/10.1017/S095382082100011X

Index